IC Array Cookbook

Walter G. Jung

HAYDEN BOOK COMPANY, INC.
Rochelle Park, New Jersey

Library of Congress Cataloging in Publication Data

Jung, Walter G.
IC array cookbook.

Includes index.
1. Integrated circuits. I. Title.
TK7874.J86 621.381'73 79-27923
ISBN 0-8104-0762-0

Printed in the United States of America

1	2	3	4	5	6	7	8	9	PRINTING
80	81	82	83	84	85	86	87	88	YEAR

Preface

Integrated circuit devices are currently available to satisfy a wide variety of applications, both digital and linear in nature. Yet, there are always application requirements that cannot be directly (or optimally) satisfied by standard, off-the-shelf IC devices. Nevertheless, these more specialized or lower-volume applications can still profit from the application of IC device characteristics through the use of *IC arrays.* An IC array, as the name implies, is simply an array of active IC circuit components—transistors of various types, diodes, and the like. Packaged in uncommitted form, such devices still retain the typical IC characteristics of high density, matched specifications, and low cost. Their outstanding attribute is their inherent flexibility—a feature that allows easy implementations of high-performance custom designs over a broad range of applications.

Once the fairly simple rules of IC circuit design are appreciated, IC arrays may be applied to achieve efficient circuit functions almost as readily as the more common IC components, such as operational amplifiers. Indeed, the special characteristics of IC array devices often allow circuits to be implemented that would be otherwise prohibitive.

This book treats the theory and practical application of IC arrays, with the emphasis on the latter. It is divided into three parts. Part I is an introduction, which covers basic IC circuit concepts and the range of IC array types available. Part II is a circuit applications section, which illustrates uses of IC arrays within a large number of circuit subgroup types, such as ac and dc amplifiers, differential amplifiers, regulator and power circuits, wideband circuits, oscillators, gain controls, and miscellaneous other uses. Part III consists of appendixes, including selected data sheets, manufacturers' source information and device cross references, and a distributor listing.

The book is written in a down-to-earth, practical style like the author's previous *IC Op Amp Cookbook, IC Timer Cookbook,* and *IC Converter Cookbook.* All circuits are presented with specified component values; a description of operation, range, and limits of operation; and suggestions for modifications and alternative uses. End-of-chapter bibliographic references will aid the reader seeking further background.

The author would like to express sincere thanks to Hal Wittlinger, Bob Rauth, and Walter Dennen of the RCA Solid State Division for their assistance in the preparation of this book. Also, general thanks go to the RCA Solid State Division for allowing the use of their technical literature within portions of this book.

My thanks also go to my wife, Anne, who once again produced the typed manuscript, and to my research assistants, Jeannie and Mark.

WALTER G. JUNG

Contents

Part I

Introduction to IC Arrays

1
IC Array Active Devices and Processes

Successful use of integrated circuit arrays starts with a good basic understanding of the active devices of which they are composed. Accordingly then, this first chapter focuses on the basic characteristics of the array elements, including npn and pnp transistors and both signal and Zener diodes. These four elements comprise the bulk of active IC array devices.

Of all the elements available in IC arrays, the npn transistor is the most widely used, and for several good reasons. First of all, the npn transistor is more easily made than any other device necessary to the junction isolation process; it therefore serves as a standard workhorse. Not only is the npn transistor more readily manufactured, its device characteristics are of a very high quality, in some regards much higher than those of analogous discrete devices.

It is quite feasible to pack a very large number of small-signal npn transistors within an IC. Although the pin limitations of array packages do not permit access to the full-density potential of the IC process, it is not uncommon to see as many as 100 transistors in a non-array IC. Most of these will be of the general-purpose, small-signal npn variety, the very types under discussion here.

Currently available IC npn transistors are capable of operation at voltages up to 40 V, currents to 100 mA, and with betas of up to 400. Where IC transistors excel, however, is in their *matching* characteristics. For instance, any two similar geometry transistors in a single IC chip will have V_{BE}'s that will typically match, within 1 mV or less, for identical operating conditions. Similarly, these two transistors will also have a gain that is matched within 10 percent (or less).

When advantage is taken of these inherent matching characteristics, IC array devices can be designed to be both efficient and economical. To do so, however, the key parameters of the transistors themselves must be understood.

NPN Transistors

Figure 1-1 illustrates a typical npn transistor in both its physical cross section and equivalent schematic. IC transistors are formed on a P-type silicon substrate, using what is termed the *junction-isolated process.* This npn transistor is composed of the n-p-n areas within the isolation well formed between the collector n region of one transistor and the next transistor collector and the substrate p region. For correct (isolated) operation, the collector region must *always* be reversed bias with respect to the substrate. This simply means that the collector must be maintained positive with respect to the substrate, that is, that the substrate must always be the lowest potential point of the circuit.

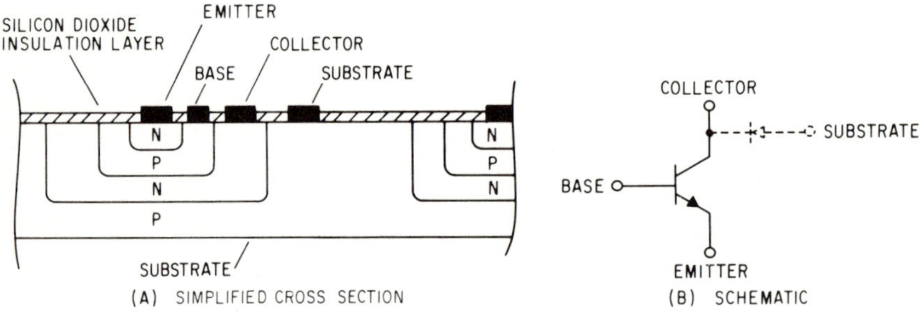

Fig. 1-1 *IC npn transistor: (a) cross section, and (b) schematic*

Since the substrate is common to all other transistors within the IC, similar bias restrictions apply to these transistors as well. A general statement that can be made about correct substrate bias is that the substrate must always be the most negative of the potentials applied to the IC if proper isolation between the circuit elements is to be maintained. A typical IC array is shown in Fig. 1-2.

Performance data typical of IC array transistor types is listed in Table 1-1. Three general types of transistors are listed here: general-purpose types as typified by the CA3046/CA3018 units, higher performance general-purpose types such as the CA3096 units, and the higher current types typified by the CA3083.

As can be noted in the table, all three transistor types have a very low offset voltage V_{IO}, indicative of good matching, and also a very low temperature coefficient of offset voltage, or low drift. The first two types

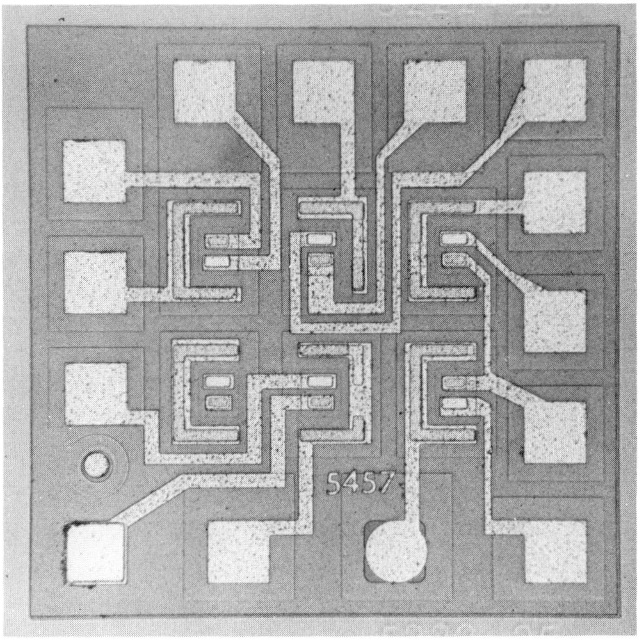

Fig. 1-2 General-purpose npn IC array CA3046 (Courtesy, *RCA Solid State Division*)

Table 1-1 *NPN Device Characteristics*

	V_{BE} at 1 mA	V_{IO} at 1 mA	TC V_{IO}	H_{FE}	F_T
General-purpose (3046/3018)	715 mV	0.5 mV	1 μV/°C	100 at 1 mA	500 MHz at 2 mA
High-performance general-purpose (3096)	690 mV	0.3 mV	1 μV/°C	390 at 1 mA	350 MHz at 4 mA
High-current (3083)	680 mV	1.2 mV	—	76 at 10 mA	—

are low-current units suitable for 10 mA or less; the third type is capable of 100 mA.

The first column of this table lists the V_{BE} for the respective transistor types at a 1-mA current level; as can be noted, all are slightly different, indicative of the different processes involved.

The general equation for the theoretical base emitter voltage of a transistor is

$$V_{BE} = n \left(\frac{KT}{q}\right) \ln \left(\frac{I_c}{I_o}\right)$$

where

K = Boltzman's constant (1.38×10^{-23})
T = absolute temperature (°K)
q = electron charge (1.6×10^{-19})
I_c = collector current
I_o = reverse saturation current (process-dependent)
n = constant (1 for npn)

A very useful point to remember is the value of KT/q at room temperature—25°C, or approximately 0.026. For the device types described here, V_{BE} at any I_c can be calculated if I_o is known. For the three types listed, I_o is as follows:

Type	I_o
3046/3018	$\approx 1 \times 10^{-15}$
3096	$\approx 2.5 \times 10^{-15}$
3083	$\approx 4.8 \times 10^{-15}$

For the 3046/3018 type, for example, the V_{BE} at a current of 100 μA at 25°C (298°K) should be

$$V_{BE} \approx 1(0.026) \ln \left(\frac{1 \times 10^{-4}}{1 \times 10^{-15}}\right)$$
$$\approx 651 \text{ mV}$$

Evaluation of typical transistor performance curves for the 3046/3018 type gives a deeper understanding of its basic behavior patterns. The data appears in Fig. 1-3.

In Fig. 1-3 (a) are shown the device input characteristics for changes in current. Examining the V_{BE} data here, it will be noted that the V_{BE} at the 100-μA current level is 651 mV, as the equation predicted. If this same curve is examined closely, it exhibits a gradual (upward) curvature above about 1 mA. This indicates a gradual deviation from theoretical behavior, or the approach of saturation.

At lower current levels—in fact, much lower than the span shown here—this transistor type will be found to follow the V_{BE} equation quite exactly. Its useful dynamic range is about six decades. While it is usable at currents over 1 mA, up to about 10 mA, the operation will no longer be theoretical, and matching will deteriorate. Evidence of this latter point is the sharp increase in V_{IO} above the 1-mA level.

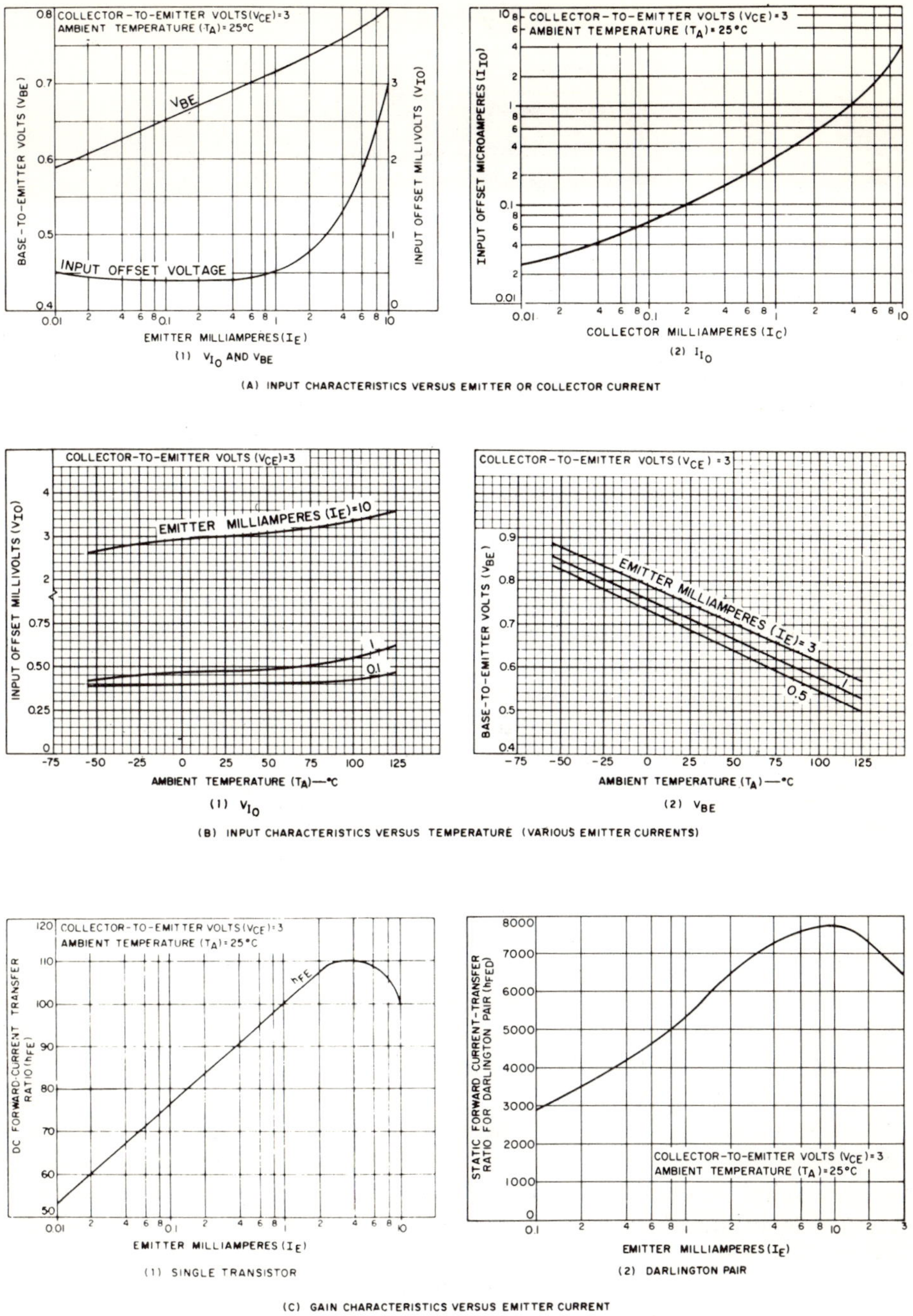

Fig. 1-3 *General-purpose npn transistors CA3046/3018: (a) input characteristics versus emitter or collector current, (b) input characteristics versus temperature (various emitter currents), (c) gain characteristics versus emitter current, and (d) f_T versus collector current*

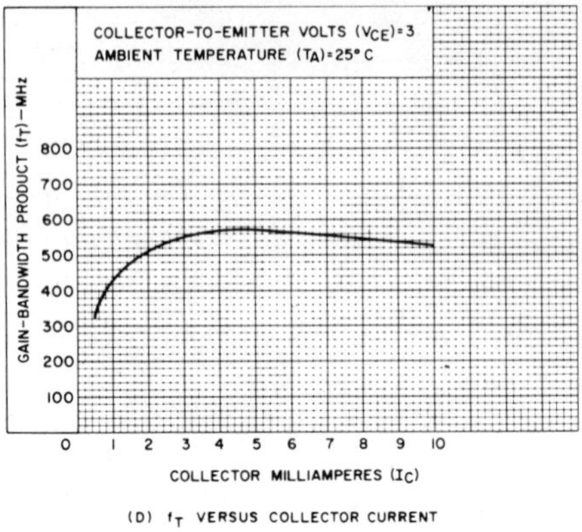

Fig. 1-3 *General-purpose npn transistors CA3046/3018 (cont'd)*

The input bias current (I_{IB}) of a transistor is that current required to flow into (or from) the base for a given emitter current. With a differential amplifier transistor pair, there are two base currents, one for each transistor of the pair.

In a balanced operating state, with each transistor of the pair conducting the same emitter currents, the base currents will be nominally equal for equal current gains (H_{FE}). Input offset current I_{IO} is a measure of a gain *mismatch* between a transistor pair (typical performance for the devices under discussion are shown). For an H_{FE} of 100 and a current of 1 mA, for example, an I_{IO} of 0.3 μA can be used to calculate the gain difference, since

$$I_{IB} = \frac{I_E}{H_{FE}}$$

An I_E of 1 mA and an H_{FE} of 100 would result in a (nominal) I_{IB} of 10 μA. The offset current of 0.3 μA tells us this current might be as high as 10.3 μA, or as low as 9.7 μA. Or, stated another way, the H_{FE} is matched $\pm$3 percent (100 $\pm$ 3).

The variation of input characteristics with changes in temperature is shown in Fig. 1-3(b). As can be noted, V_{IO} changes are very low for low currents but become substantial at high currents. At the 100-μA current level, the change in V_{IO} is less than 1 μV/°C, a very respectable figure.

V_{BE} change versus temperature is shown in Fig. 1-3(b) for various current levels. From this figure, the actual V_{BE} temperature coefficient can be read. For example, at 1 mA, V_{BE} changes -320 mV over the 175°C range from -50 to $+125$°C, or -1.83 mV/°C. The exact V_{BE} temperature coefficient (TC) changes with emitter current, as can be noted from the different slopes for the three curves.

The gain characteristics of this device for various current levels are illustrated in Fig. 1-3(c). A single transistor and a Darlington paired combination are shown. Note that over the ranges given the change in H_{FE} is only about 2/1. H_{FE} falls off above 10 mA in both cases.

NPN IC transistors such as these are capable of very high speed because of their low capacity and high f_T. Above a current level of about 1 mA, the f_T reaches several hundred MHz, as shown in Fig. 1-3(d).

High-Performance, General-Purpose NPN Types

The transistor process used in the CA3096 device features much improved performance over that of the more pedestrian 3046/3018 units. The 3096 type has an H_{FE} that is nearly four times higher, and even more closely controlled matched characteristics. In addition, the breakdown voltage capability is increased to 30 V. Figure 1-4 illustrates the general performance of this type of device.

Input characteristics, including V_{IO} and V_{BE} versus collector current, are shown in Fig. 1-4(a). As can be noted from the V_{IO} data, V_{IO} is very low, typically 300 μV at currents between 100 μA and 1 mA, increasing above 1 mA. V_{BE} characteristics of this device versus current are shown in Fig. 1-4(a) and versus temperature in Fig. 1-4(b). The latter clearly shows how the temperature coefficient (TC) of V_{BE} changes for different currents.

The excellent gain characteristics of this type of device are shown in Fig. 1-4(c). Gain is nearly 400 at 1 mA and holds up well as high as 10 mA, as well as at low currents.

High-Current NPN Device Characteristics

The device type of those available in array form with higher current capability is illustrated in Fig. 1-5. The characteristics are typical of the CA3083 transistor types.

Input characteristics of the device type are shown in Fig. 1-5(a); as can be noted, both V_{IO} and I_{IO} are higher than the corresponding characteristics of the device types in Figs. 1-3 and 1-4. However, in the V_{BE} characteristic, it can be noted that the curve extends to 10 mA without deviating from its essentially straight-line form. This indicates that adherence to the theoretical model is better at higher levels of current, that is, up to 10 mA. This conformity is due to the larger geometry of this device type compared to that of the others.

Gain characteristics of this device are shown in Fig. 1-5(b). As can be noted, the gain is less than that in Figs. 1-3 and 1-4 at low levels but is superior at levels above 10 mA. This property makes this transistor type suited to driver or buffer applications.

The basic device type (3083) is usable at voltages as high as 15 V; the 3183 is a higher-voltage version, with a 30-V rating.

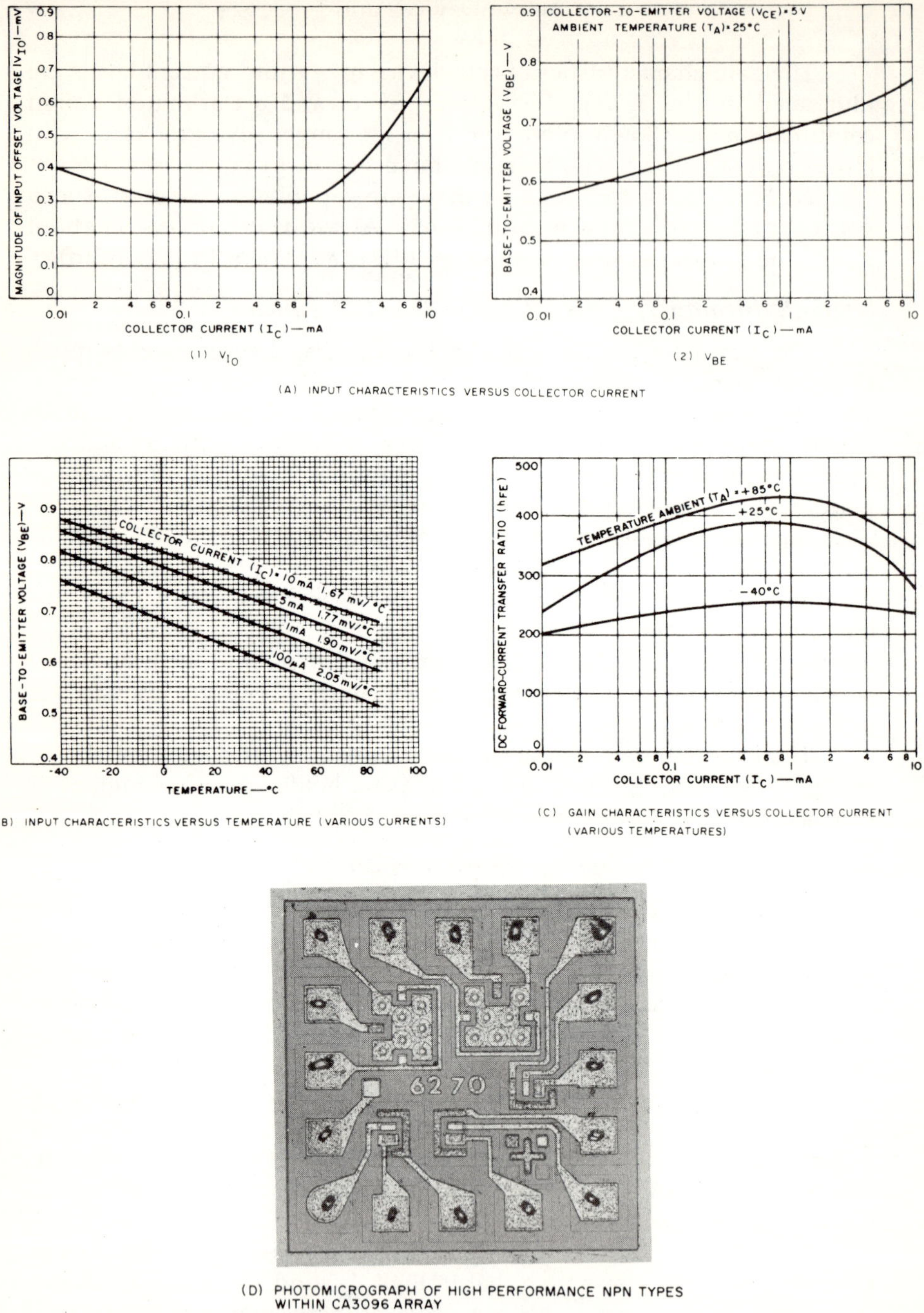

Fig. 1-4 *High-performance npn transistors CA3096: (a) input characteristics versus collector current, (b) input characteristics versus temperature (various currents), (c) gain characteristics versus collector current*

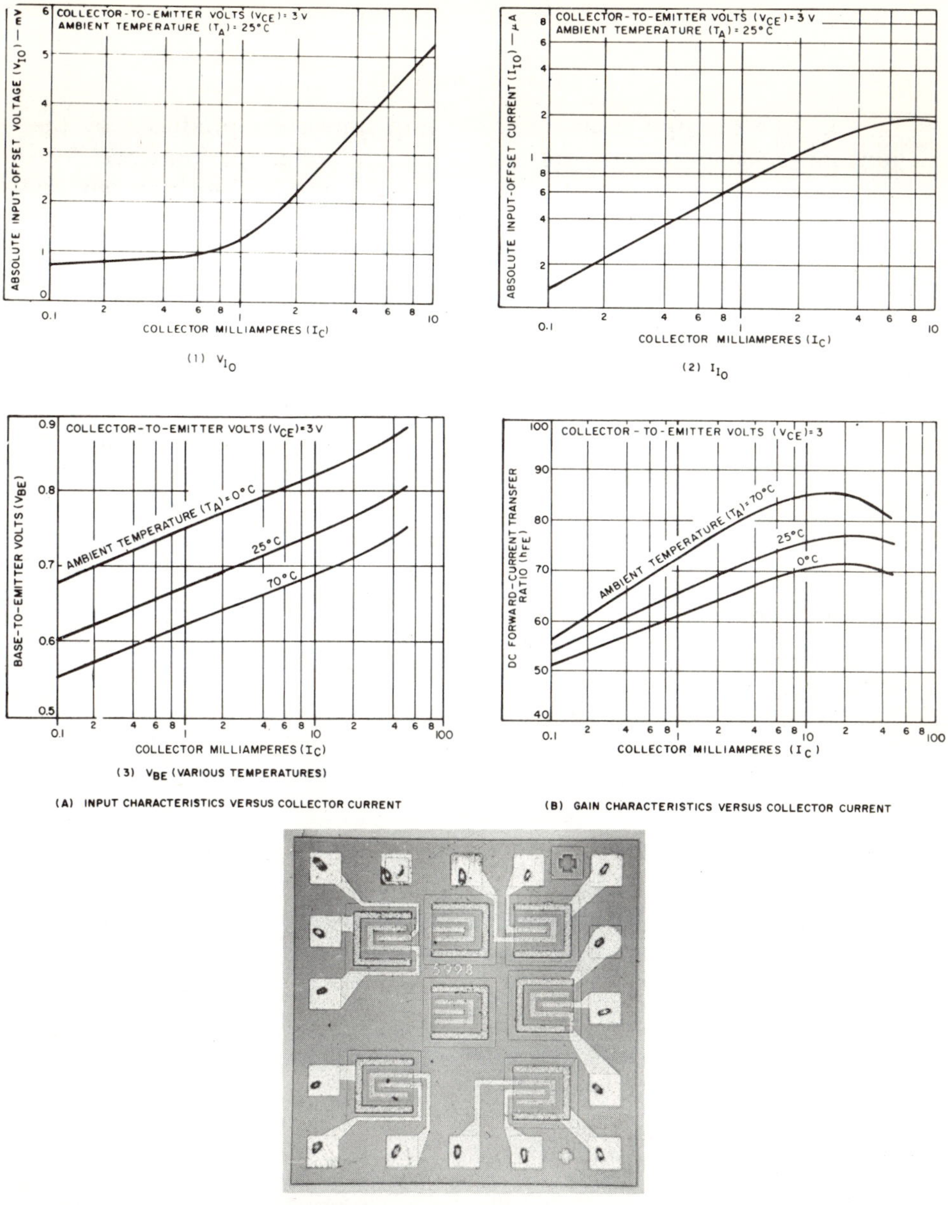

Fig. 1-5 *High-current npn transistors CA3083: (a) input characteristics versus collector current, (b) gain characteristics versus collector current, and (c) photomicrograph of high-current npn type array CA3083* (Courtesy, *RCA Solid State Division*)

(various temperatures), and (d) photomicrograph of high-performance npn types within CA3036 array (Courtesy, *RCA Solid State Division*)

PNP Transistors

Unlike the npn transistors, the pnp transistors produced by the junction-isolated process are somewhat limited in performance. The reason is that the manufacturing process is optimized for npn devices but not for pnp devices.

Figure 1-6(a) illustrates why this is so, showing a simplified cross section of an IC pnp transistor. A pnp transistor is actually formed as a modification of the normal npn process. In the drawing, the n region acts as the base of the pnp transistor but is actually what would normally be a collector in an npn device. Into this n material are diffused p-type collector and emitter regions, as shown. The overly large base region of this pnp transistor severely limits its frequency response, and the typical f_T of a device formed by this technique is 4 to 5 MHz, far less than for the npn. The gain is also low, typically 40, and useful gains are limited to collector currents of 1 mA or less.

As can be noted from the structure, the current flow of this pnp is

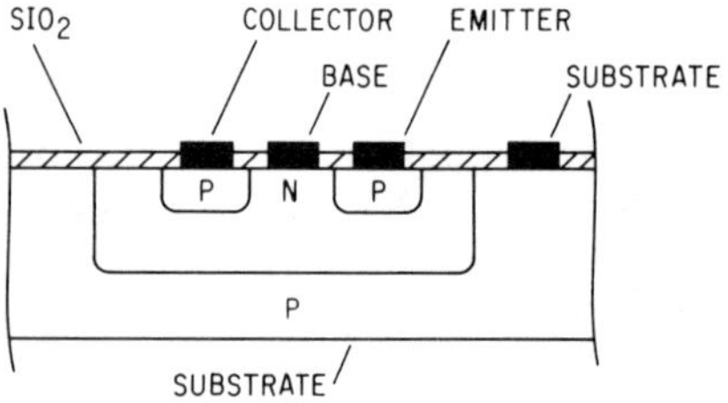

(A) SIMPLIFIED CROSS SECTION

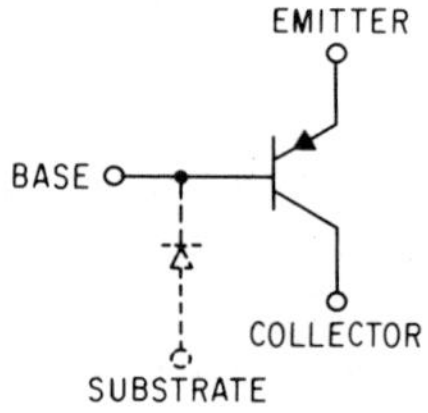

(B) SCHEMATIC

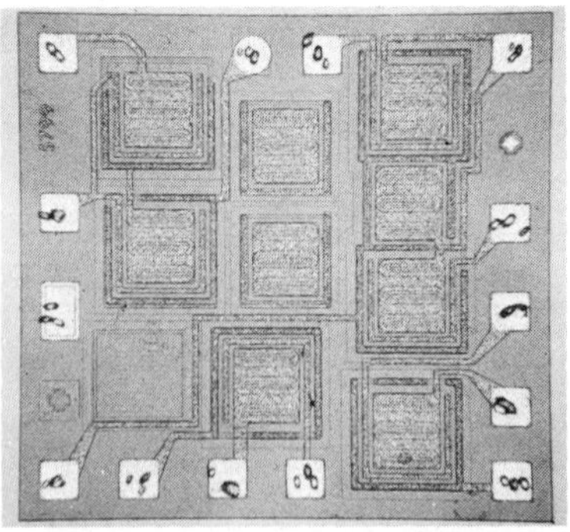

(C) PHOTOMICROGRAPH OF LATERAL PNP TRANSISTORS (CA3084)

Fig. 1-6 *IC pnp transistor: (a) cross section, (b) schematic, and (c) photomicrograph of lateral pnp transistor CA3084* (Courtesy, *RCA Solid State Division*)

Table 1-2 PNP Device

	V_{BE} at 100 μA	V_{IO} at 100 μA	TC V_{IO}	H_{FE}
General-purpose (3084)	590 mV	0.4 mV	0.5 $\mu V/°C$	40 at 100 μA
High-performance general-purpose (3096)	600 mV	0.15 mV	0.5 $\mu V/°C$	85 at 100 μA

lateral, thus the name, *lateral pnp.* A very important point also is that there is *parasitic* transistor action associated with this scheme; in Fig. 1-6(a), for instance, there is a *substrate* pnp transistor formed between emitter, base, and substrate, as well as a collector base substrate. In the actual device process used, the gain of these parasitic devices must be minimized by process control so as to best optimize the desired lateral characteristics. Geometrically, lateral pnp's are usually laid out in a circular or square pattern to optimize the gain, as shown in the photograph of a typical lateral pnp in Fig. 1-6(c).

An IC array device with characteristics similar to those being discussed is the CA3084. The typical electrical characteristics of this general-purpose lateral pnp device are summarized in Table 1-2.

The V_{BE} of a lateral pnp follows the same general form as that of the npn, or

$$V_{BE} = n \left(\frac{KT}{q}\right) \ln \left(\frac{I_c}{I_0}\right)$$

In this case, however, the constant n is approximately 1.3. For the device type discussed here, I_o is 1.5×10^{-12}.

One notable (and desirable) property of these pnp's that does *not* degrade is their matching characteristic. As can be noted in the table, both device types have V_{IO}'s below 0.5 mV at 100 μA, with the improved-performance 3096 device at 0.15 mV. The V_{IO} and V_{BE} characteristics of the 3084 device type are shown in Fig. 1-7(a). As can be noted from the V_{BE} curve, the line has a curvature at about 1 mA, indicating that it is best used at lower levels when V_{BE} accuracy is important.

The V_{IO} and V_{BE} characteristics versus temperature of the 3084 device are indicated in Fig. 1-7(b). V_{IO} shows a relatively small drift with temperature, as can be noted.

Gain characteristics of this device are illustrated in Fig. 1-7(c). The H_{FE} of a single device is shown to be relatively low. Gain is a maximum

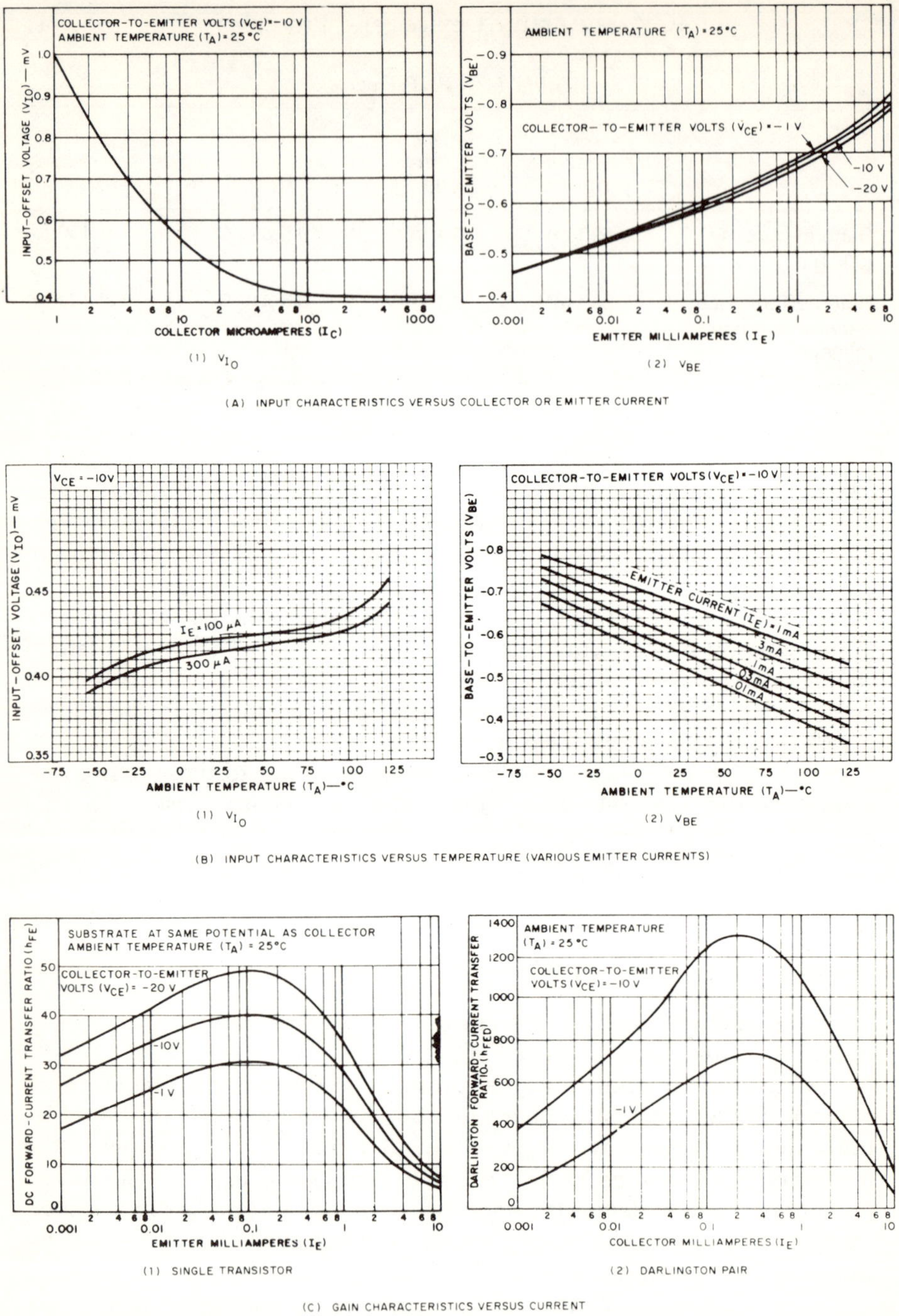

Fig. 1-7 General-purpose lateral pnp transistor CA3084: (a) input characteristics versus collector or emitter current, (b) input characteristics versus temperature (various emitter currents), and (c) gain characteristics versus current

of 40 at a 100-μA operating current; it falls off rapidly at higher or lower levels and also varies with voltage.

A substantial increase in the useful gain of this device can be had by operating two of its transistors in a Darlington mode, as shown in Fig. 1-7(c). It will be noted that gain peaks at around 1000 at a curent of 100 to 200 μA.

Higher Performance PNP

In the CA3096 device type, an improved process is used, and this improvement is clearly reflected in the performance, as indicated in Table 1-2. Matching is better, gain is higher, and the device is much more useful, generally speaking. Device characteristics are shown in Fig. 1-8.

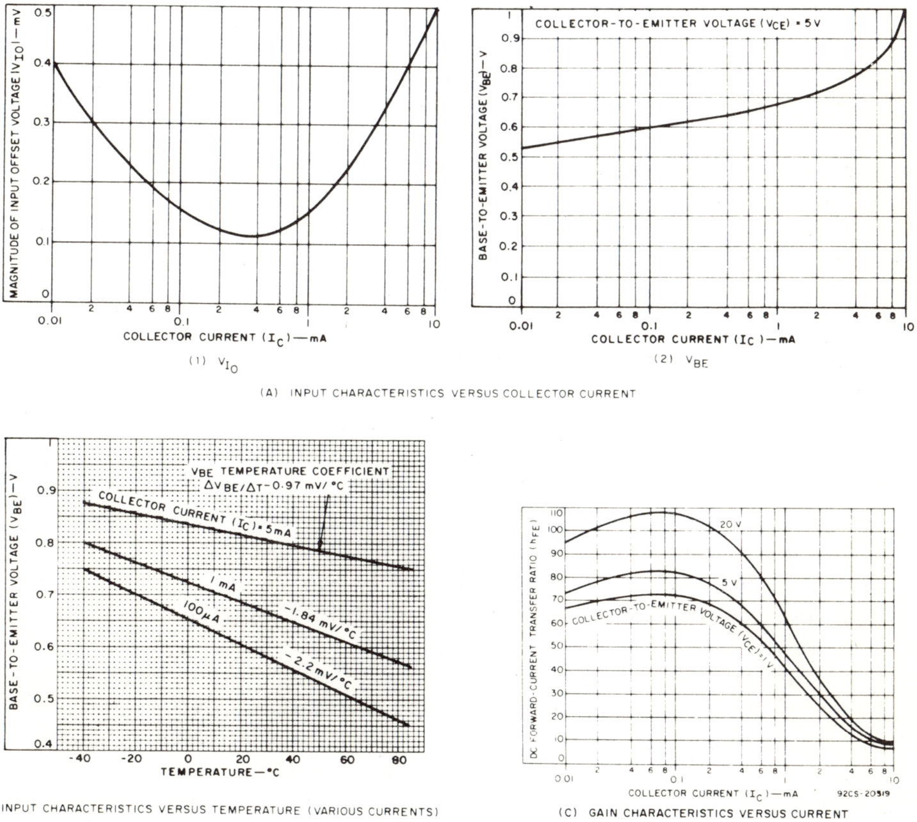

Fig. 1-8 High-performance lateral pnp transistor CA3096: (a) input characteristics versus collector current, (b) input characteristics versus temperature (various currents), and (c) gain characteristics versus current

The input characteristics shown in Fig. 1-8(a) indicate a very low V_{IO} and V_{BE}; they hold up well versus current and with less sensitivity to collector voltage than exhibited in other devices. The V_{BE} TC is shown in Fig. 1-8(b).

Perhaps the most notable single performance of this device is the gain characteristic, as shown in Fig. 1-8(c). Gain is not only higher at the current where it reaches a maximum, but it also does not fall off quite as rapidly for higher or lower current levels.

A common characteristic of both lateral pnp device processes is a relatively high breakdown voltage, as compared to that of general-purpose npn's. Both device processes have typical V_{CE} ratings of 40 V.

Diodes

In the IC junction-isolated process, two basic diode connections are the most often used; they will be called here simply "low-voltage" and "high-voltage" diodes. They are shown in Fig. 1-9, and their characteristics are summarized in Table 1-3.

In a low-voltage diode, an npn transistor base emitter junction is used, with the collector shorted to the base, as shown in Fig. 1-9(b). In making the diode, base-collector contacts are simply metallized together for use as the anode, and the emitter becomes the cathode. This diode's forward characteristics are nearly ideal, and the forward-voltage characteristic may be expressed in the same manner as the V of the npn transistor. The similar desirable properties of matching voltages hold for these diodes as well as for the general-purpose transistors, as they are in fact made from the same transistor process.

Since the reverse characteristics of this diode are those of the transistor EB junction, reverse voltages must be limited to about 6 V or less to prevent Zener action, which occurs at about 7 V. The two basic diode device types derived from this process are the CA3039 and CA3019. They are specified somewhat differently for reverse leakage (Table 1-3). Both types feature low reverse capacity of 1 to 2pF or less, which enables high speed application. Their speed results from the fact that the transistor is never allowed to saturate.

The forward-voltage characteristics of such a diode-connected transistor are very similar to the V_{BE}–I_C characteristics of the analogous npn transistor, as can be noted in Fig. 1-10(a). Diode-matching characteristics (V_{F_1}–V_{F_2}) are also similar in behavior to the V_{IO} of an npn pair, as shown in Fig. 1-10(a). Diode forward characteristics versus temperature are shown in Fig. 1-10(b), and again similarity is noted. The offset voltage, also shown in Fig. 1-10(b), is low for low forward currents but increases for higher current levels, when the diode is no longer operating near the theoretical model.

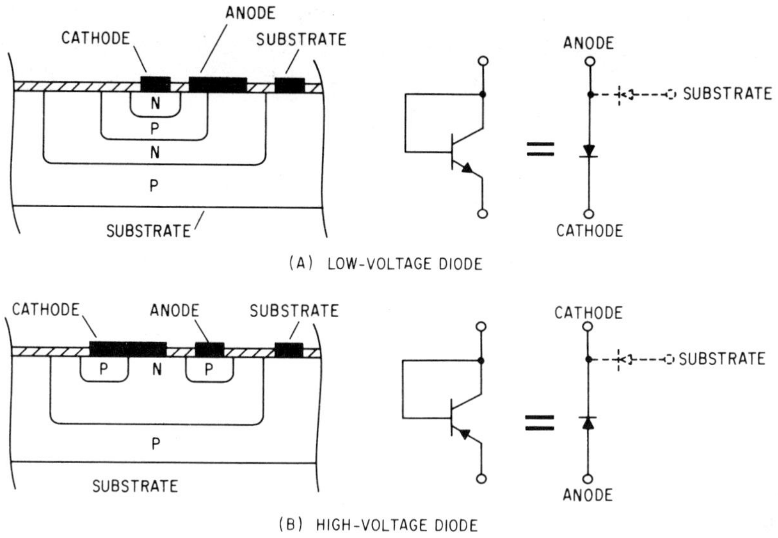

Fig. 1-9 *Low-voltage and high-voltage diodes: (a) low voltage, and (b) high voltage*

Table 1-3 *Diode Characteristics*

	V_F at 1 mA	V_{F1}–V_{F2} at 1 mA	TC V_F at 1 mA	I_R	C_R
Low-voltage (3039/	730 mV	0.5 mV	−1.9 mV/°C	16 pA at 4 V	<1 pF
3019)	730 mV	1 mV	—	5 nA at 4 V	<2 pF
High-voltage (3141)	780 mV	0.6 mV	−1.5 mV/°C	100 nA (MAX) at 20 V	<1 pF

The reverse-leakage characteristics of a low-voltage E–B diode are shown in Fig. 1-10(c) for varying temperatures and a 4-V bias. At room temperature, the reverse leakage is very low—on the order of 20 pA.

The dynamic resistance of a diode operated in the small-signal region with relatively low current changes is given as

$$r_D = \frac{KT}{I_F q}$$

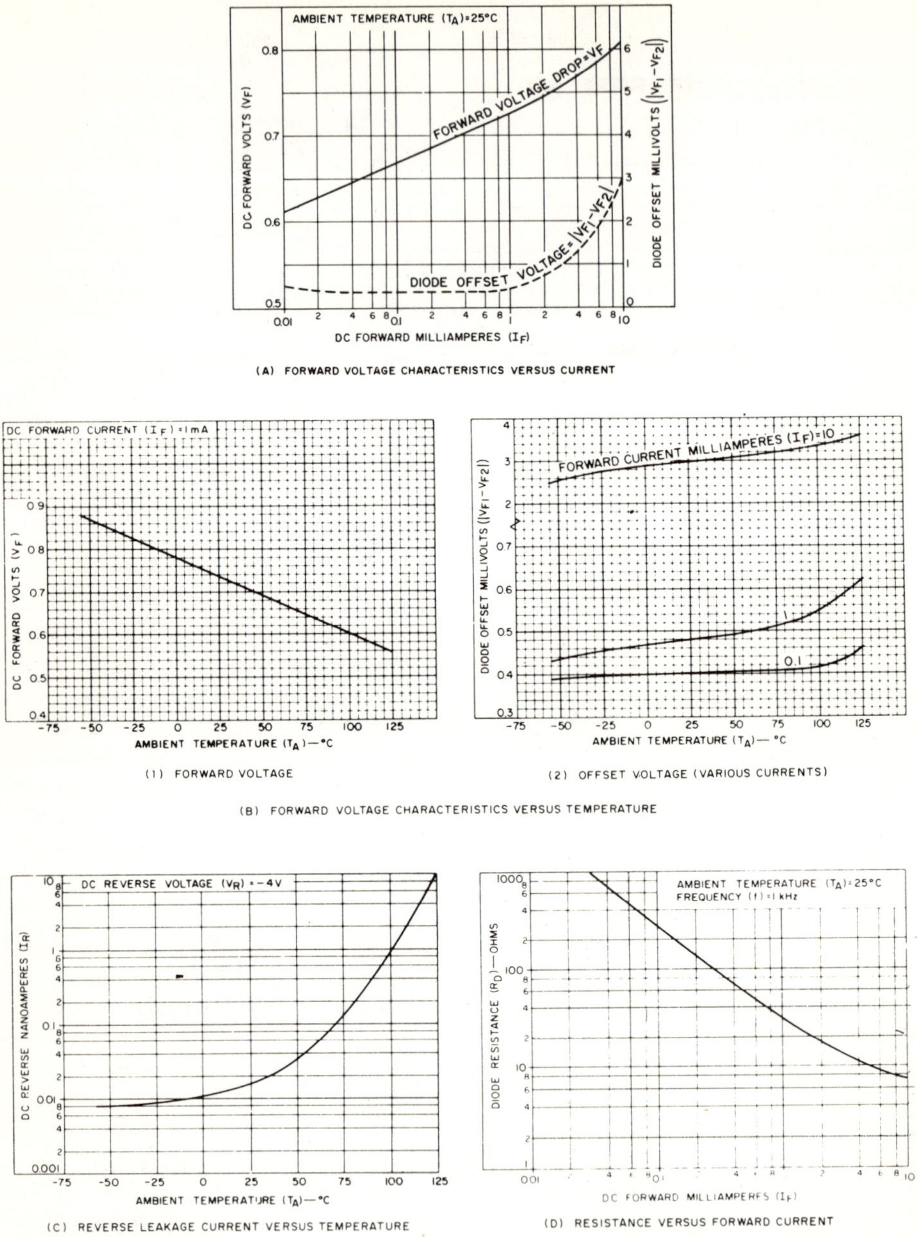

Fig. 1-10 *Diode characteristics of the CA3039: (a) forward voltage characteristics versus current, (b) forward voltage characteristics versus temperature, (c) reverse leakage current versus temperature, and (d) resistance versus forward current*

where I_F is the forward current (dc). The easily remembered rule of thumb is that this resistance is 26 Ω (at room temperature) for a 1-mA current. The dynamic resistance of a low-voltage diode is shown in Fig. 1-10(d) and it follows this equation closely for currents below 1 mA.

High-voltage diodes are obtained in the junction-isolated process by altering a pnp transistor, as shown in Fig. 1-9(a). The base collector junction of a pnp transistor is shorted, and the emitter is used as the anode, thus producing a reverse breakdown characteristic similar to the V_{CE} of the pnp, or about 40 V typically.

Unfortunately, this connection also forms an inherent parasitic pnp between the anode of the diode and the substrate, with the cathode as the parasitic base. Therefore, a current will flow through this transistor, as well as from anode to cathode. This parasitic action is intrinsic to this type of diode, but the shunt (parasitic) current can be controlled to a degree (although not eliminated). In the CA3141, a device type using this process, the parasitic current is typically 4 percent of the desired anode current.

Zener Diodes

The Zener diode is the fourth major active device type available in IC arrays; it is a natural fallout from the basic npn-device EB junction.

A reverse junction characteristic with a Zener property is obtained in the npn transistor, breakdown occurring at around 7 V. A Zener is formed by using the emitter-base junction as a Zener, with the collector strapped to the emitter (to minimize the possibility of parasitic action). This is shown in Figs. 1-11(a) and (b). The collector should not be forward-biased or left floating, as substrate conduction is not desired.

The typical characteristics of Zener diodes such as these are shown in Table 1-4. The two basic types are the CA3093 device, which is specifically characterized for Zener applications, and the general-purpose EB junction from such units as the 3046 or 3018 family. The latter types are not specified for Zener service but are often so used.

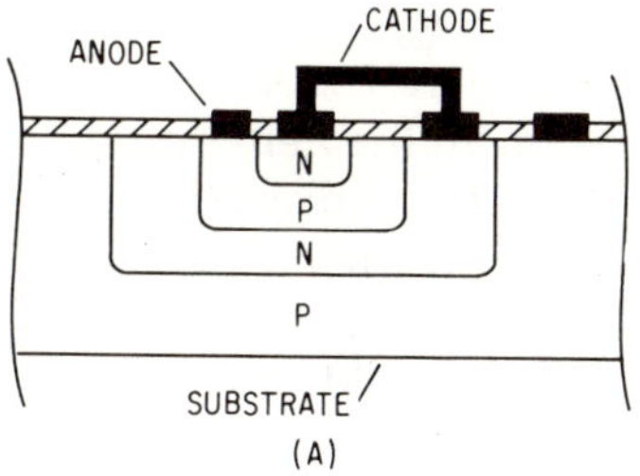

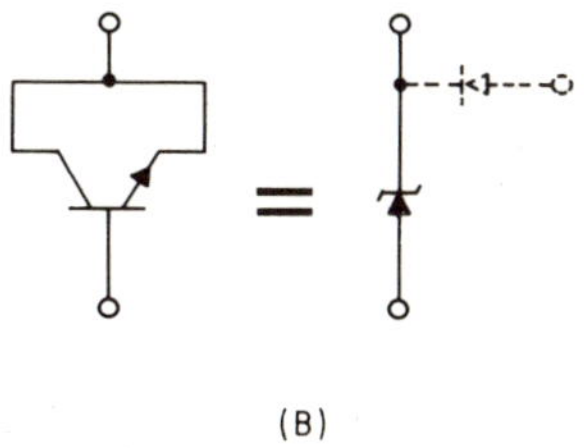

Fig. 1-11 *Zener diode: (a) cross section, and (b) schematic*

Table 1-4 Zener Diode Characteristics

	V_Z	V_{Z1}–V_{Z2}	TC V_Z	R_Z	I_Z(MAX)
High-current types (3093)	7 V at 10 mA	200 mV	≃+3.6 mV/°C (0.05%/°C)	15 Ω at 10 mA	35 mA
General-purpose E-B (3046/3018)	7 V at 100 μA	200 mV	≃+2 mV/°C (0.03%/°C)	100 Ω at 1 mA	35 mA

Zener voltage is uniform in the different junctions of a given chip, just as V_{BE} is repeatable. A pair of Zeners connected back-to-back, for instance, have a uniform breakdown or symmetry, a difficult-to-match function in discrete types.

The TC of the EB junction is positive and varies with current. For these processes, the VZ will vary in the range of 2 to 3.5 mV/°C, which is equivalent to 0.05 percent/°C or less. Refer to Fig. 1-12(a).

Dynamic resistance is high in the smaller general-purpose transistors, typically about 100 Ω at 1 mA. It is appreciably lower in the larger geometry types, about 15 Ω for the same current level [Fig. 1-12(b)]. Zener current rating is listed as 35 mA (max.) for both types, limited by power dissipation.

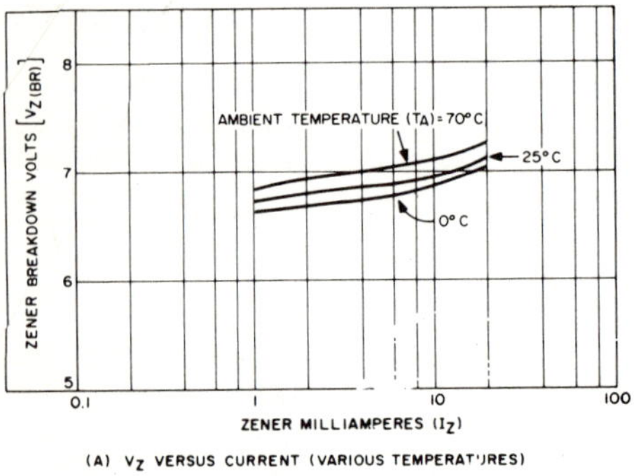

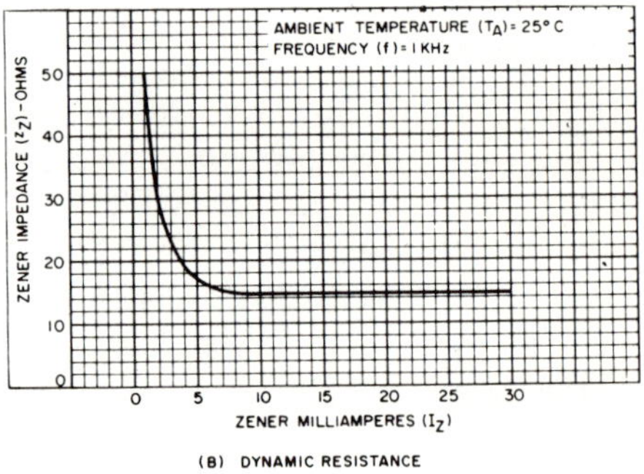

Fig. 1-12 Zener characteristics (CA3093): (a) V_Z versus current (various temperatures), and (b) dynamic resistance

A point of interest in Zener use of standard transistors (such as the general performance types) is that operation in the Zener mode *degrades* the noise and gain of the transistor.[1] Therefore transistors subjected to reverse Zener currents should not be considered for subsequent amplifier uses. Also, devices connected as amplifiers should be carefully *protected* against possible EB voltages, which could cause Zener action. Protection of differential amplifiers against excessive input overvoltage is a case in point. Protection can consist of a simple pair of back-to-back signal diodes to limit V_{IN} to $\pm V_F$.

[1] See Motchenbacher, C. D., Ref. 9.

REFERENCES FOR CHAPTER 1

(1) Grebene, A. B., *Analog Integrated Circuit Design,* Van Nostrand Reinhold, 1972.

(2) Camenzind, H. R., *Electronic Integrated Systems Design,* Van Nostrand Reinhold, 1972.

(3) ———, *Monochip Designers Handbook,* Interdesign, Sunnyvale, Calif., 94086.

(4) ———, *XR-Chip Custom IC Design Kit Handbook,* Exar Integrated Systems, Sunnyvale, Calif., 94086.

(5) Leidich, A. J., "Linear Monolithic Circuit Techniques," paper # 740012, Society of Automotive Engineers, Inc., 1974.

(6) Hoover, M. V., "Monolithic Linear IC Arrays and Their Applications," Proceedings of *EEE* magazine IC Seminar, March-April, 1969.

(7) Camenzind, H. R., Grebene, A. B., "An Outline of Design Techniques for Linear Integrated Circuits," *IEEE JSSC,* June 1969.

(8) Device Data sheets CA3046, CA3018, CA3096, CA3083, CA3084, CA3039, CA3019, and CA3093 from RCA Solid State Division, Route 202, Somerville, N.J., 08876.

(9) Motchenbacker, C. D., "Protect Your Transistors Against Turn on or Transient Damage," *Electronics,* Dec. 6, 1971.

(10) Jung, W. G., "Inside Linear ICs—Transistor Arrays and You," *Radio Electronics,* Nov. 1971.

(11) Jung, W. G., "Inside Linear ICs—Practical Applications of Transistor Arrays," *Radio Electronics,* Dec. 1971.

(12) Jung, W. G., "7 Ways to Use IC Transistor Arrays", *Radio Electronics,* Jan. 1972.

Technical data for this chapter was supplied through the courtesy of RCA Solid State Division.

2
Array Devices and Design Techniques

In this chapter, we will take a closer look at the IC array circuit elements and how they are interconnected to form basic circuit subsections. While IC array active devices can certainly be utilized with design techniques generally suitable for discrete devices, *maximum* advantage is gained when they are utilized in a manner that capitalizes on their inherent features. Design techniques that do this are generally those of the monolithic IC art and have ground rules different from those for discrete techniques.

While IC array devices have some limitations, the freedom and liberties they provide are many. Once the circuit designer has mastered the slightly different approach necessary, he will find using them highly advantageous.

Of course, designing with arrays is not *exactly* the same as designing a monolithic IC, as many circuit elements (resistors, etc.) are still external to the IC. Also, the finite pin limits of arrays necessarily restrict the number of transistors available per package. Nevertheless, for those special applications not satisfied precisely by a standard committed type, IC arrays provide decided advantages.

The discussions within these pages take the reader step by step through circuit design techniques optimum for array use. Not only will the ideas be useful in working with arrays, they will greatly increase the understanding of virtually all linear ICs, since they are all based on similar circuitry.

Current Sources

Regulated sources of current, or simply "current sources" as they are most often called, are a fundamental element of linear ICs. Where a source (or sink) current is required to be stable with regard to applied voltage and/or loading, current source techniques are useful to the designer. Within ICs, they are used to stabilize bias levels, as level shifters, and to control power dissipation.

With the matched devices available in arrays, a wide variety of current sources (and sinks) can be implemented easily. Often, the use of a current source will enhance the overall operation of a circuit appreciably. While this statement may not be obvious at first, it will become more apparent as a variety of circuits are discussed.

Conventional Current Source

A simple, constant current source that is familiar to discrete circuit designers is shown in Fig. 2-1. In Fig. 2-1(a) the base of Q1 is biased at a fixed potential by the resistor divider R1–R2 across the voltage V+. If V+ is stable, the voltage at Q1 will be stable; hence the voltage across R3 (I_{R3}) will be stable. Therefore, the output current I is stable. The circuit will be sensitive to the V_{BE} drift of Q1 unless the divider voltage is selected to provide a high voltage across R3, which makes V_{BE} effects relatively small. The circuit can handle high currents (> 10 mA) if a composite connection is used for Q1, such as a Darlington (to be discussed). As shown, the circuit sinks current. A source of current can be implemented with a pnp for Q1, and a reversal of supply voltage.

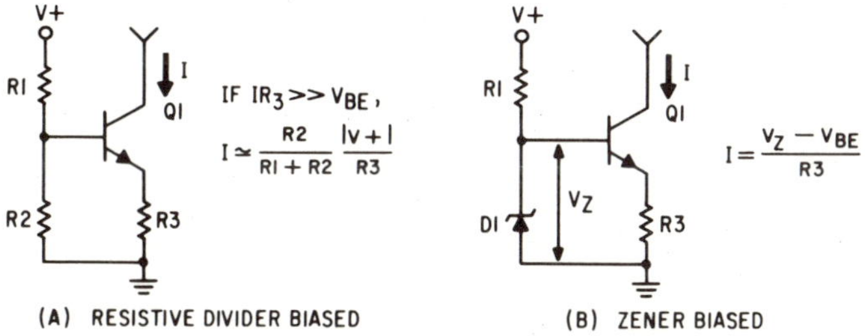

Fig. 2-1 Conventional current source: (a) resistor-divider-biased, and (b) Zener-biased

The weak point of this circuit is its direct dependence upon a regulated voltage for V+ and three stable resistors. The modification shown in Fig. 2-1(b) removes this limitation. Here, R2 is replaced with a Zener diode, which applies a voltage V_Z to the base of Q1. The voltage across R3 (and I) is now dependent only upon V_Z and V_{BE}. The output current is therefore dependent upon V_Z, V_{BE}, and R_3, as shown.

This is quite a useful circuit, as Zener diodes exist in every monolithic transistor E–B junction. With the proper voltage selection for V_Z, the TC of the output current can be made to approach zero. Modifi-

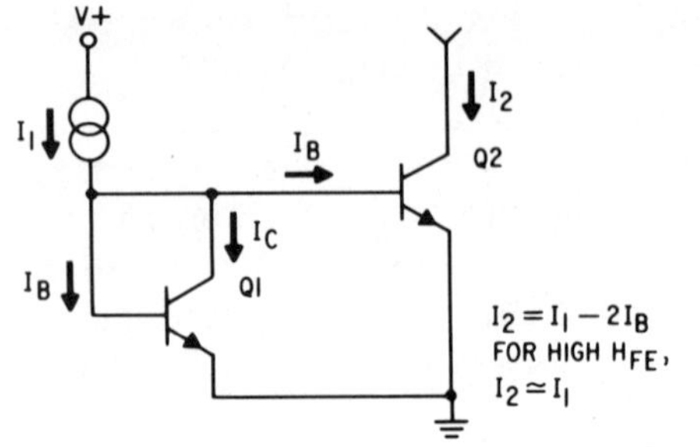

Fig. 2-2 *Basic monolithic current source*

cations of the circuit are possible, as described for the simple resistor-biased circuit.

Monolithically Matched Current Sources

The most common (and most simple) current source used in monolithic circuits is shown in Fig. 2-2.[1] In this circuit, Q1 and Q2 are a pair of matched monolithic transistors with similar V_{BE} and current gain. The circuit is also called a "diode-biased" current source because of Q1's diode connection.

The current I_1 is a reference current, and the output current I_2 is made to be related to it. Thus, if I_1 is stable, I_2 will be stable. In actuality, this circuit is a *referenced* or slave current source that duplicates (mirrors) I_1. Therefore, it is also known as a "current mirror" circuit.

If Q1 and Q2 are well matched, Q1 will bias itself at some V_{BE} related to current I_1. Q2, being matched to Q1, will then conduct a similar current since its base is in parallel with the V_{BE} of Q1. Thus, I_2 is made to be equal to I_1.

I_2 is not *exactly* equal to I_1, however, because the reference current I_1 loses 2 I_B (to Q1 and Q2). Thus, I_2 (IC of Q2) is equal to I_1 less the base currents, as shown. If Q1 and Q2 have high H_{FE} (100 or more), this error will only be 2 percent or less.

This simple current source is a workhouse circuit in ICs and has many variations. I_1 can be set up by a simple resistor to V+, and I_2 will then be proportional to V+. If I_2 is to be independent of supply voltage, a current regulator circuit [similar to that in Fig. 2-1(b)] is required for I_1. This most basic circuit can be turned around to work for pnp transistors as well.

A scaled-current output version of the circuit is shown in Fig. 2-3. Here, Q1 is connected as previously, but two transistors, Q2 and Q3, are paralleled for the output. With Q1, Q2, and Q3 all identical, the output

[1] See R. J. Widlar, Ref. 1.

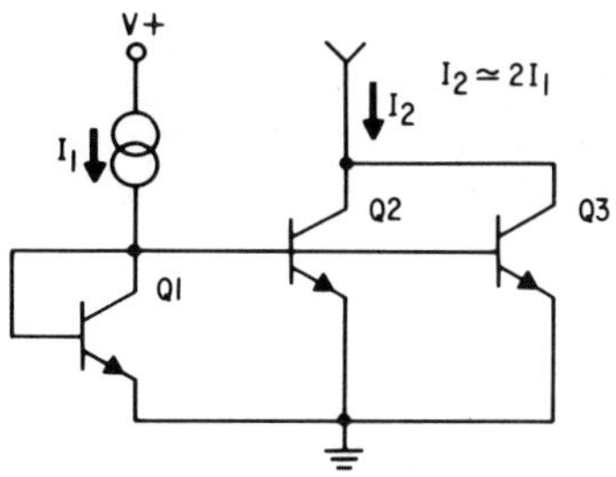

Fig. 2-3 *Scaled current source*

I_2 equals (approximately) 2 I_1. The circuit is subject to base-current errors, as mentioned previously. PNP versions are possible as well, as was true for Fig. 2-2.

The basic current source of Fig. 2-2 can be improved in terms of its output impedance by the addition of some emitter resistance, as shown in Fig. 2-4. This allows the current I_2 to be more independent of voltage changes occurring at the collector of Q2.

Resistors R1 and R2 are selected to drop a small (200-mV or less) voltage with currents I_1 and I_2 flowing. When R1 equals R2, the currents will be nominally equal; changes in current "gain" both above and below unity are possible by altering the ratio, as shown.

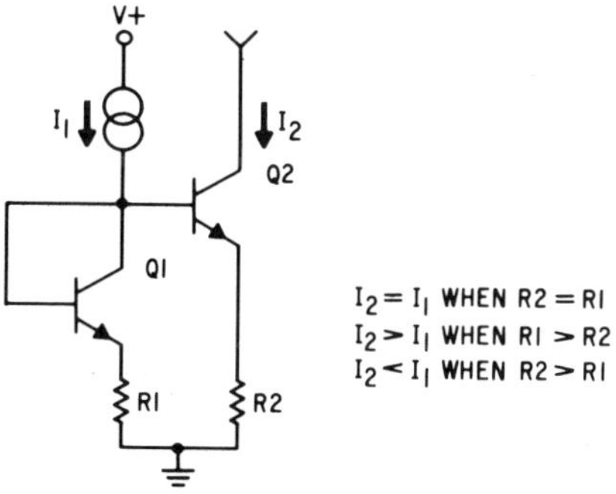

Fig. 2-4 *Current source with local feedback*

The Wilson current source, shown in Fig. 2-5, improves markedly on the simple matched transistor current source in two regards.[2] The first of these is the fact that it is much less sensitive to voltage changes at the collector of Q2, or, in other words, its output impedance is higher. This makes it much more useful in critical circuitry. The second improvement relates to current match. Whereas the basic current source

[2] See G. R. Wilson, Ref. 7.

has a 2-percent error for an H_{FE} of 100, the Wilson current source has an error of about 0.02 percent for the same transistor gain. As a consequence, it is much more useful in precisely balanced circuits or with transistors with low beta. The latter point makes this configuration most useful with pnp IC transistors, which typically have much lower beta than their npn counterparts. A pnp version of the Wilson current source is shown in Fig. 2-5(b). The pnp transistors function similar to the npn devices in Fig. 2-5(a).

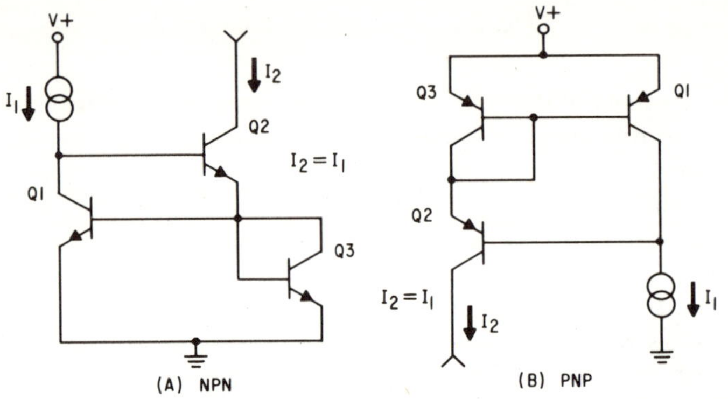

Fig. 2-5 Wilson current source: (a) npn, and (b) pnp

Both npn and pnp versions of this current source can be modified, with the variations described for the basic version. Inclusion of emitter resistance, for example, allows trim of balance or alteration of symmetry.

Logarithmic Current Source

The circuit of Fig. 2-6(a) is a useful one for the generation of low-valued currents; it does not depend on the high-valued resistors usually associated with low currents. This circuit operates upon the principle that the difference in emitter base voltage between matched transistors at different currents is very predictable. This voltage, or V_{BE} difference, is

$$\Delta V_{BE} = \left(\frac{KT}{q}\right) \ln \frac{I_1}{I_2}$$

where I_1 and I_2 are the currents in the two transistors.

In Fig. 2-6(a), Q1 operates at current I_1 and Q2 at a lower current, I_2. Q2 will exhibit a lower V_{BE} than Q1 because of this lower current. The difference in the V_{BE}'s of Q1 and Q2 appears across R_A and determines the output current. Thus, there is another name for this current source, the "ΔV_{BE}" current source, since the V_{BE} difference is actually the reference voltage appearing across R_A.

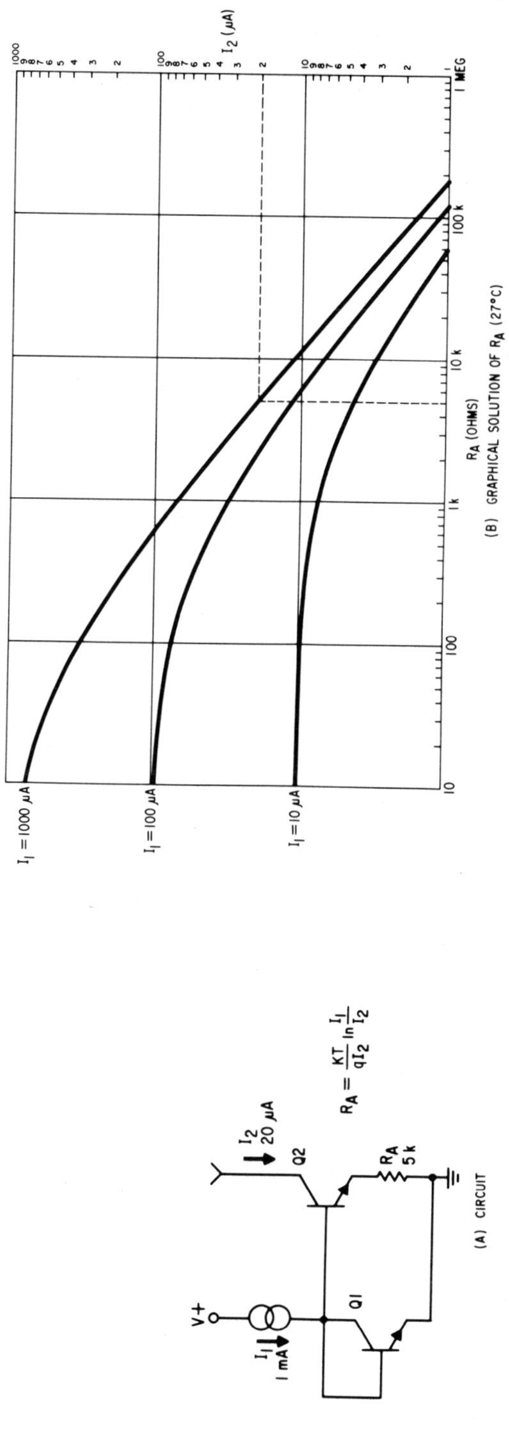

Fig. 2-6 *Logarithmic current source: (a) circuit, and (b) graphical solution of* R_A, where, at 27°C (300°K), $R_A = (KT/qI_2)\ln(I_1/I_2)$

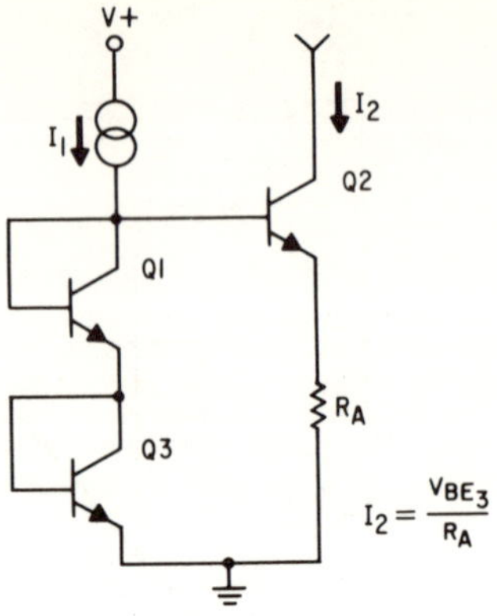

Fig. 2-7 *V_{BE}-referenced current source*

In operation, the selection of R_A for a given value of I_2 is aided by a nomograph, such as that shown in Fig. 2-6(b). In this example, the reference current I_1 is 1 mA, and I_2 is desired to be 20 μA. A horizontal line is drawn at the I_2 level of 20 μA until it intersects the $I_1 = 1000$ μA curve. The value of R_A is they read as 5000 Ω (bottom). Curves for reference currents of 100 and 10 μA are also included. Note that this circuit can also be used for pnp transistors as well, to which the nomograph is equally applicable. Other graphical solutions to this type of circuit problem are described in the references.

Figure 2.7 illustrates a circuit that can be used to produce a temperature-dependent current. In this configuration, the voltage reference across R_A is the V_{BE} of Q3 since the V_{BE}'s of Q1 and Q2 essentially cancel.

This circuit is useful when temperature compensation is needed, in which case a negative coefficient for I_2 is required. I_2 decreases in the same proportion as V_{BE} with increasing temperature, or about 0.3%/°C (TC $\simeq -0.3$%/°C).

Buffered Current Source

The buffered current source of Fig. 2-8 reduces the errors of the simple diode-biased current source of Fig. 2-2, particularly when multiple output currents are required.

In Fig. 2-8(a), npn transistors Q1 and Q2 are matched to one another, and buffer transistor QB is added to provide their base currents, with minimal error in I_1. Note that since QB supplies only *base currents* to Q1, Q2, etc., the error in I_1 is very small since the base current of Q1–Q2 is reduced again by H_{FE}. The error between I_1 and I_2 can be 0.1 percent or less. Additional stages can be added, such as Q3 and Q4 (or more), without appreciably increasing errors. Scaling of I_2, I_3, I_4, etc., is also possible.

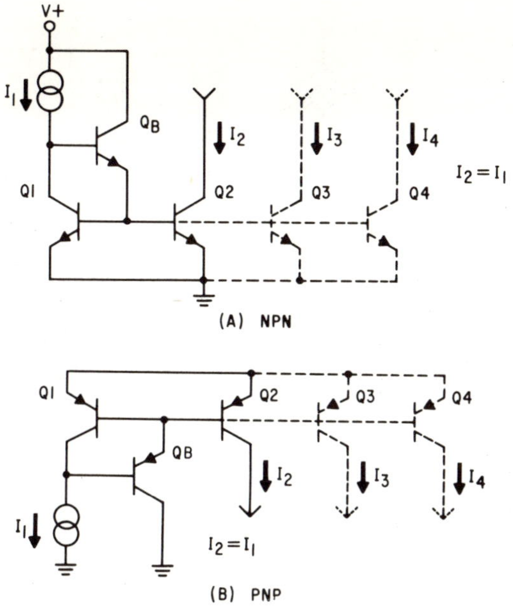

Fig. 2-8 Buffered current source: (a) npn, and (b) pnp

The pnp version shown in Fig. 2-8(b) is a preferred form of current source (or current mirror) for pnp's when close tracking or multiple outputs is desired.

For both of these configurations, modifications such as emitter resistor feedback and logarithmic scaling can be utilized as well, as described above. For example, individual emitter resistors applied to Q2, Q3, etc., can allow outputs over a wide dynamic range.

Voltage References

Another basic circuit element in ICs is a stable voltage reference. Voltage reference circuitry takes on a wide variety of forms within ICs, depending upon the exact requirements. However, many voltage references are based upon Zener diodes, since they are readily available as E–B junctions. Other forms of voltage references are band-gap circuits and, for modest requirements, simple forward-diode stacks. The following discussion surveys the various techniques.

Basic Zener-Diode Regulators

The simplest form of Zener regulator is shown in Fig. 2-9(a). Here, an emitter base junction is fed from a voltage V+ that is greater than 7 V, setting up a Zener current I_Z. If V+ is relatively stable, the current source shown can be replaced by a resistor. This configuration is ade-

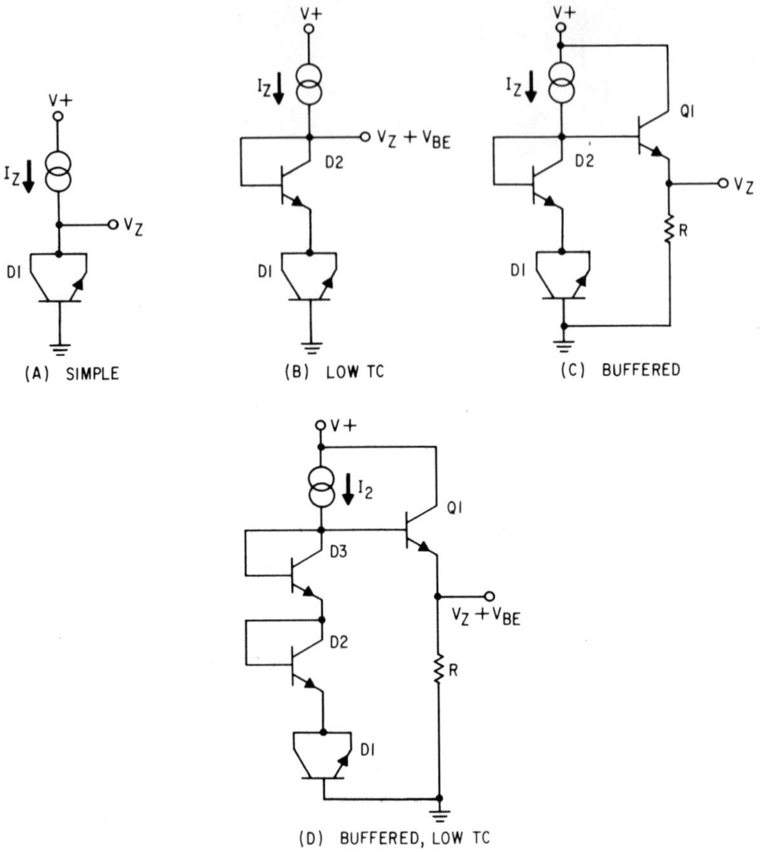

Fig. 2-9 *Basic Zener regulators: (a) simple, (b) low TC, (c) buffered, and (d) buffered, low TC*

quate when the output load is relatively light or the change in load is relatively small.

The TC of this simple regulator will be characteristic of the junction used, typically about 2 to 3 mV/°C (positive) for array devices. With a Zener voltage of ≃7 V, this equates to +0.03–0.04%/°C.

A lower TC can be achieved with the circuit of Fig. 2-9(b), which adds a forward-biased diode junction in series with the Zener. This roughly cancels the positive TC of the diode. If TC is to be minimum, the Zener current I_Z should be optimized to a level when the positive and negative TCs are equal. I_Z should be regulated to minimize current changes and loading light (or constant). This is because changes in I_Z result in Zener voltage changes.

Figure 2-9(c) shows how an emitter follower (Q1) can be added to

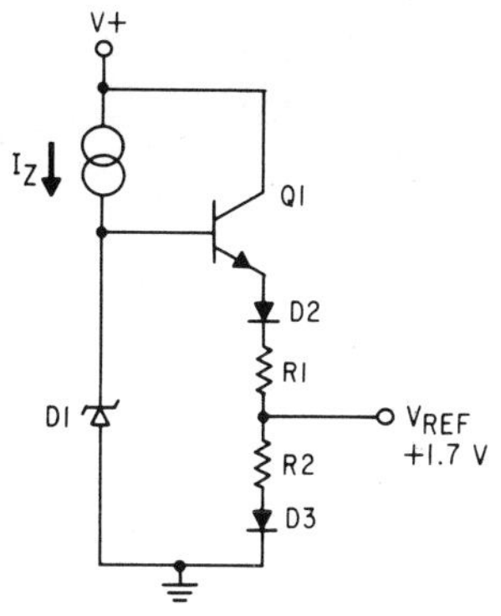

Fig. 2-10 *Low-voltage reference*

minimize loading effects on V_Z. As shown in this case, the V_{BE}'s of Q1 and D2 will cancel, leaving the output voltage V_Z and the basic TC of Zener D1.

To restore a low TC, the buffered low TC connection of Fig. 2-9(d) can be used. D2 cancels the positive TC of D1, whereas the negative TCs of D3 and Q1 cancel one another. The output is at a level of $V_Z + V_{BE}$, and I_Z should be regulated for best performance.

The methods just described are often not optimum in terms of obtaining a zero TC voltage inasmuch as the required equal and opposite junctions TCs are not attained. Figure 2-10 illustrates a method of producing a regulated reference voltage that can (in theory) be adjusted by resistor ratios for minimum TC.

In this circuit, Zener diode D1 is biased by current I_Z and exhibits a positive TC.[3] Q1 buffers the Zener voltage and sets up a current in the string, D2–D3–R1–R2. The forward voltages of Q1 and D2 increase the positive TC of the Zener while reducing the net voltage. On the other hand, diode D3 provides a negative TC. When R1 and R2 are selected in an appropriate ratio, these positive and negative TCs cancel, and the output V_{REF} has a nominally zero TC. For typical monolithic devices, V_{REF} will be zero TC in the range of 1.7 V. This circuit can be seen in such IC voltage regulators as the LM300, LM305, and others.

Start-Up Current Regulator

For applications in which the Zener voltage must be maintained closely in the presence of wide variations in input voltage, a circuit similar to Fig. 2-11 can be used. This circuit uses a current regulator consisting of Q1, Q2, and Q3 to bias the Zener at the desired current, I_Z, which is set up by R_E. The current regulator configuration of Q2–Q3

---

[3] See R. J. Widlar, Ref. (11).

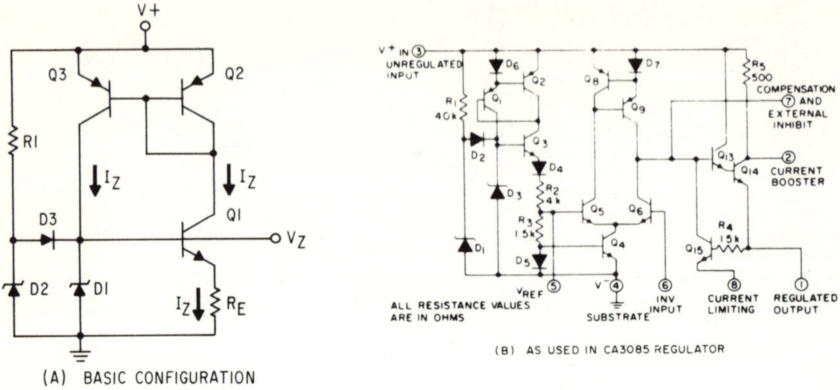

Fig. 2-11 *Start-up current regulator: (a) basic configuration, and (b) as used in CA3085 regulator* (Courtesy, *RCA Solid State Division*)

rejects input-voltage variations by 60 dB or more, allowing V_Z to be maintained constant over a wide range and without dissipation increases at high V+ levels.

The circuit is *not* self-starting, as the Zener actually biases itself into conduction. It is started by the resistor Zener diode network, R1–D2–D3. Once Zener D1 is fired, D3 becomes back-biased, and variations in R1–D2 current are not seen in V_Z.

This circuit is shown here in relatively simplified form; variations mentioned previously can be used, such as an OTC Zener connection, buffering, etc. The circuit is used in various ICs, among them the CA3085 voltage regulator, as shown in Fig. 2-11(b). As will be noted, the CA3085 also uses the low-voltage OTC reference scheme and employs a Wilson pnp current mirror in the I_Z current regulator.

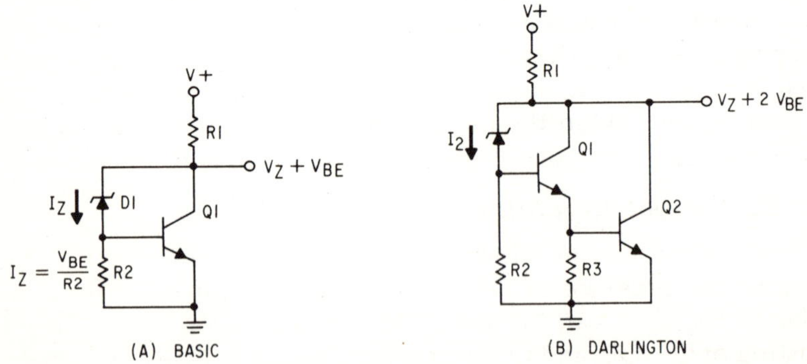

Fig. 2-12 *Amplified Zener shunt regulators: (a) basic, and (b) Darlington*

"Amplified" Zener Regulators

Another means of increasing the performance of a Zener diode regulator is placing the Zener itself in a negative feedback loop; this stabilizes the Zener current and increases overall current-handling capacity. This type of regulator is illustrated in Fig. 2-12.

In Fig. 2-12(a) is shown a basic form of this regulator with Zener D1 connected as a negative-feedback path around Q1. This circuit will bias Q1 at a collector voltage of $V_Z + V_{BE}$. The current I_Z is set by the V_{BE} of Q1 as shown and is relatively stable with current load changes. The bulk of the load current is handled by Q1.

With proper selection of operating current I_Z, the TC of this connection can be quite low; the connection is thus suitable for use as a voltage reference. An example of the use of this circuit can be found in the μA723 voltage regulator IC.[4]

The Darlington version of the circuit, shown in Fig. 2-12(b) decreases further the variation in I_Z that results from load current changes. Note that this circuit can handle currents up to 10 mA or more. In both circuits, the requirement for a stable input current feed is not present since the very low dynamic impedance provided by feedback absorbs current variations without large output voltage changes.

Diode Stack Regulator

There are often requirements for simplified forms of voltage regulators that can be satisfactorily handled by simple circuitry. One of these is the diode stack regulator shown in Fig. 2-13.

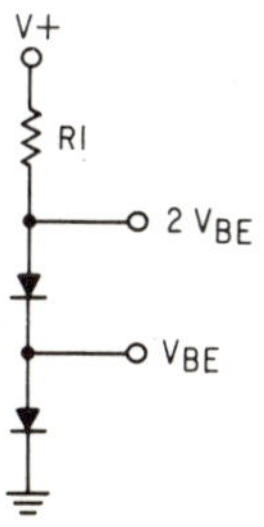

Fig. 2-13 *Diode stack regulator*

In this circuit, a series of two (or more) diodes is stacked in a series string, with the junction taps providing various multiples of V_{BE} reference voltages. While the voltage at the taps will vary with temperature, this may not always be a disadvantage.

---

[4] See Lieux and Ricks, Ref. (12).

The scheme provides excellent decoupling of line noise because of the low diode impedances.

V_{BE} *Multipler*

The V_{BE} multiplier of Fig. 2-14 is another type of regulator that provides a voltage related to V_{BE}, using a transistor as an amplifier. The base emitter voltage is multiplied by the gain factor, which is (R1 + R2)/R1.

Note that this circuit can produce any required multiple of V_{BE} (not just *integer* multiples). The output voltage is available at a low impedance level. This circuit is often used for the biasing of class B output stages in which a temperature-tracked V_{BE} voltage is required.

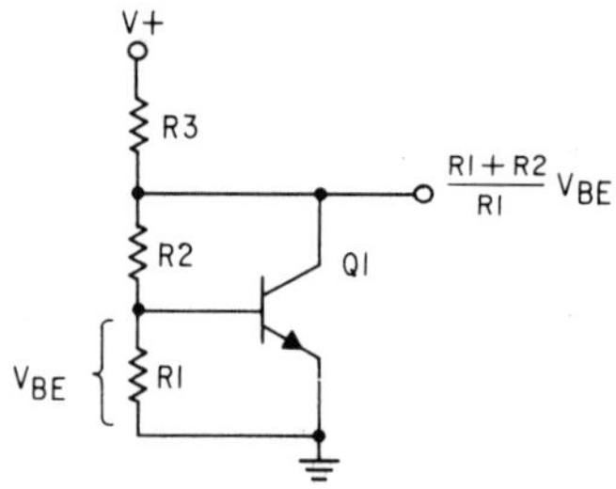

Fig. 2-14 V_{BE} *multiplier*

Bandgap Voltage References

A more recent form of reference voltage regulator is the "bandgap" voltage reference, so named because of its being based on the bandgap voltage of silicon. The basic advantage of the bandgap voltage reference as opposed to the more traditional Zener diode is that it is a *low-voltage* reference as a result of the fact that the bandgap voltage for silicon is 1.205 V. Consequently, bandgap references operate from very low supply potentials—as low as 2 V—as opposed to a Zener reference's minimum voltage of about 8 V. Another advantage of the bandgap reference is that it is inherently more quiet than a Zener diode, which is notoriously noisy.

Simplified circuits for two representative bandgap voltage types are shown in Fig. 2-15. In Fig. 2-15(a) is shown a diagram of the basic circuit concept for the LM309 voltage regulator.[5] In this circuit, diode-connected Q1 and transistor Q2 form a logarithmic current source, as previously discussed. Q2 operates at a much lower current than Q1; therefore, the difference in their base emitter voltages appears across R3, or ΔV_{BE}, where

---

[5] See R. J. Widlar, Ref. (13).

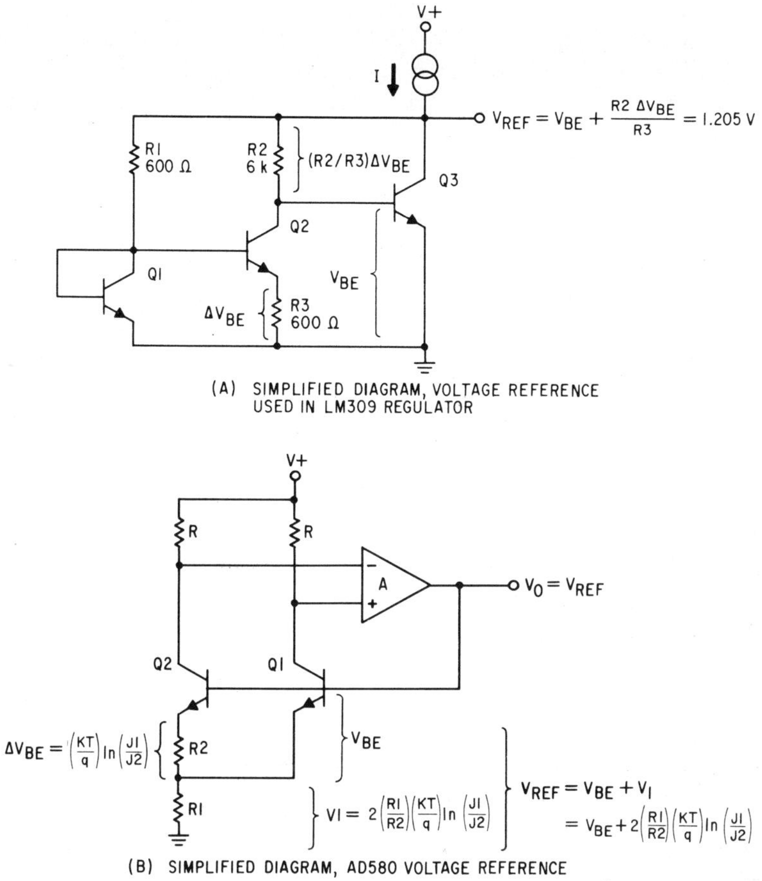

Fig. 2-15 *Bandgap voltage references: (a) LM309 voltage regulator, and (b) AD580 voltage regulator*

$$\Delta V_{BE} = \left(\frac{KT}{q}\right) \ln \frac{I_2}{I_1}$$

The voltage across R2 is the amplified form of ΔV_{BE}, or (R2/R3) ΔV_{BE}. As can be noted from the circuit configuration, this voltage and the V_{BE} of Q3 add up to the output voltage, V_{REF}.

The ΔV_{BE} component has a positive TC, whereas V_{BE} has a negative TC. When these two voltages add up to the bandgap voltage, 1.205 V, zero TC operation results. The actual circuit implementation used in the LM309 is much more complex and is discussed in more detail in the reference cited.

Another IC bandgap reference circuit is the LM313, a two-terminal

1.22-V reference diode.[6] Yet a third is the AD580, the basic circuit diagram of which is shown in Fig. 2-15(b).[7] The AD580 is a 2.5-V output voltage reference IC, with TCs of as low as 10 ppM/°C. The circuit used is different from that in Fig. 2-15(a) and employs a pair of transistors in a differential mode. In this circuit, the ΔV_{BE} appears across the emitter resistor of Q2, R2. Q1 and Q2 operate at the same current, but Q2 has an emitter area eight times that of Q1; therefore, its emitter current density is one-eight that of Q1. This creates the ΔV_{BE} component across R2 and ultimately across R1, according to the relationship shown. The regulated voltage, which is the sum of V_{BE} and the ΔV_{BE} component, appears at the bases of Q1–Q2. In this simplified circuit, $V_o = V_{REF}$, but, in the actual device, a voltage divider between V_o and Q1–Q2 scales V_{REF} to higher levels. Further information on the AD580 (and related devices) is contained in the references listed.

Level Shifters

The requirement of a level shift circuit usually is seen in a multistage dc amplifier where both input and output dc levels have specific limits or ranges, as shown in Fig. 2-16.

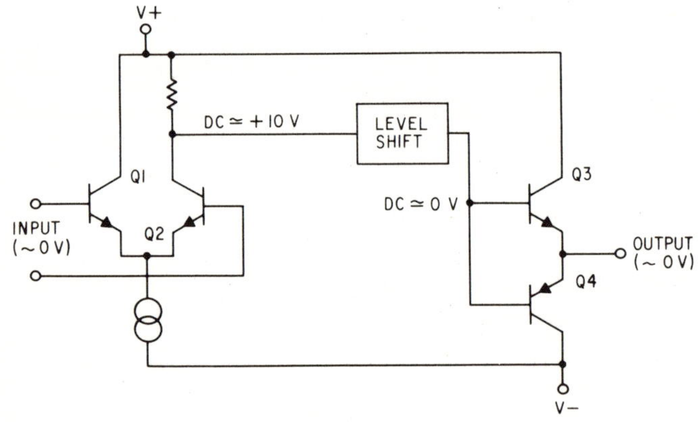

Fig. 2-16 Level shifter with dc amplifier

In this circuit, which is a (much simplified) dc amplifier, the gain stage Q1–Q2 has an input common mode level that is quiescent at 0 V. As a necessity, the output collector of Q2 will bias quiescently at some higher voltage, say +10 V. Then, if the output is to be at a quiescent

[6] See R. C. Dobkin, Ref. (14).
[7] See A. P. Brokaw, Ref. (15).

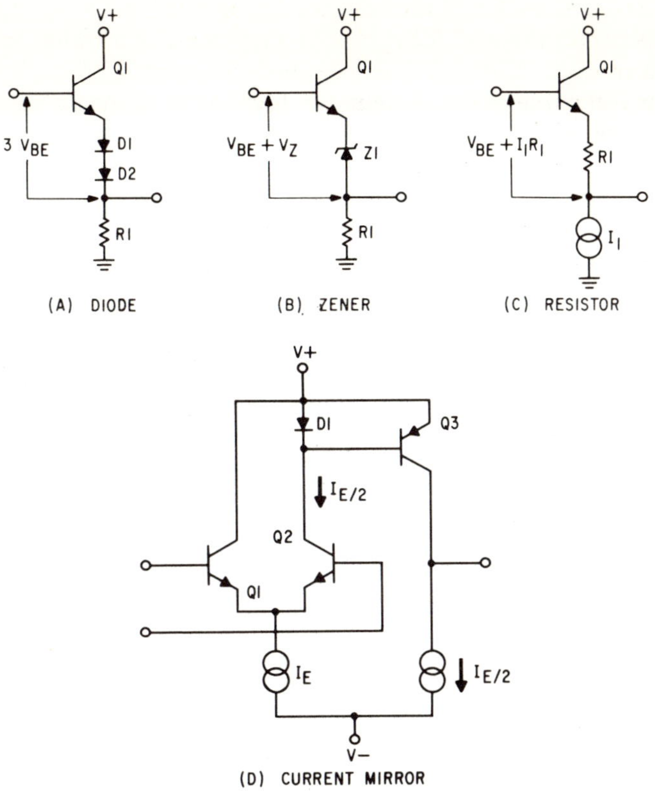

Fig. 2-17 *Level shift circuits: (a) diode, (b) Zener, (c) resistor, and (d) current mirror*

level of 0 V dc, there must be a dc level shift of −10 V between Q2 and Q3–Q4. This function is performed by a level shifter.

The general requirement for a level shifter is that it provide a fixed dc offset, but it should introduce no signal loss (such as a resistive divider). Several examples of level shift circuits are shown in Fig. 2-17.

In Fig. 2-17(a) is shown a simple diode level shifter, which shifts by a level of 3 V_{BE}. More (or less) level shift can be accommodated by altering the number of diodes in the string.

The Zener level shift circuit of Fig. 2-17(b) shifts by an amount $V_{BE} + V_Z$ and is generally more flexible. In some circuits, R1 may be replaced by a fixed current source to increase flexibility and reduce dissipation. The Zener may require by-passing to minimize noise in this circuit if at a low-level point.

The resistor-current source circuit of Fig. 2-17(c) is even more flexible, as it can be designed for any desired dc level shift simply by choosing R1 and I_1. Unlike that of the first two circuits, the output impedance of this one is high (value of R1); thus, it may require a buffer to prevent loading losses.

A very flexible arrangement for level shift purposes is the pnp current mirror scheme of Fig. 2-17(d). In this circuit, the collector current of Q2 is mirrored by D1–Q3, and Q3 is loaded by the current source, $I_E/2$. The full current output of Q2 is available at the collector of Q3, but the *voltage* at this node can be any potential between V+ and V− (minus transistor saturation drops).

Whereas this circuit is most flexible in terms of dc level shifting, it is somewhat limited in frequency response with typical monolithic pnp's. One of the first three circuits is preferred if wide bandwidth is required.

Basic Amplifiers

A pair of basic common-emitter amplifier stages are shown in Fig. 2-18, along with their governing equations of behavior. The stages are, of course, the fundamental ac gain blocks, so that an understanding of them lays the groundwork for other circuits to follow.

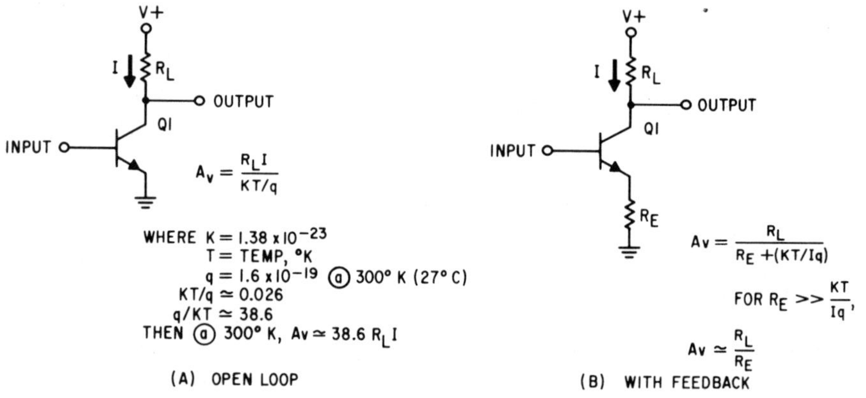

Fig. 2-18 *Basic amplifier stage: (a) open loop, and (b) with feedback*

In Fig. 2-18(a) is shown the simplest type of common-emitter amplifier, a transistor Q1 biased at a current I and loaded by R_L. This stage will have a small signal gain that is governed by R and I, as shown, with K and q as constants and T as absolute temperature. Thus, the basic gain is proportional to R_L and I and inversely proportional to temperature, the equation being

$$A_V = \frac{R_L I}{\left(\frac{KT}{q}\right)}$$

At room temperature (300°K), q/KT is equal to approximately 38.6, and a simplified expression can be used to calculate gain:

$$A_V \simeq 38.6\ R_L I$$

Note that gain is a *nonlinear* function of input voltage, and low distortion results only for *small* input voltages (10 mV or less).

In practice, gain stages such as have been described are often used with feedback when linearity is required. A simple example of a gain stage with local feedback for the purpose of linearization and gain stabilization is shown in Fig. 2-18(b).

In this circuit, an emitter resistor, R_E, is added that effectively increases the emitter impedance. In so doing, it increases input impedance, lowers gain, and improves both small signal linearity and gain/temperature stability.

Note that the gain equation for this circuit includes terms for both R_E and KT/Iq. When R_E is 10 or more times KT/Iq, the gain can be approximated by the simplified expression. R_E also reduces the circuit's temperature sensitivity, of course, as well as its linearization.

Biasing

Biasing of an IC transistor gain stage is illustrated in Fig. 2-19. In Fig. 2-19(a) is shown a bias circuit very close to what is typically used in discrete circuits, where R1 and R2 set up a base voltage and R_E establishes the desired emitter current. This method can also be used for arrays, of course, with the added availability of a matched diode that compensates for the temperature sensitivity of Q1 and D1. Such operation will result in quite stable dc bias levels.

A more efficient way to bias a monolithic transistor is to take advantage of its matching parameters, as shown in Figs. 2-19(b), (c), and (d). These circuits are all variations on the matched diode scheme discussed under current sources.

In Fig. 2-19(b), Q1 is biased at a current I, and matched device Q2 is desired to be biased at this current also. A small-value base resistor, R_B (selected for less than a 5-mV drop), will also force Q2 to conduct a current I, since the V_{BE}'s of Q1–Q2 are essentially the same. A load resistor R/2 then causes the collector of Q2 to rest quiescently at $\frac{1}{2}$ V+, an optimum class-A level.

The disadvantage to this circuit is the low value necessary for R_B, which reduces input impedance. The stage can be modified to use

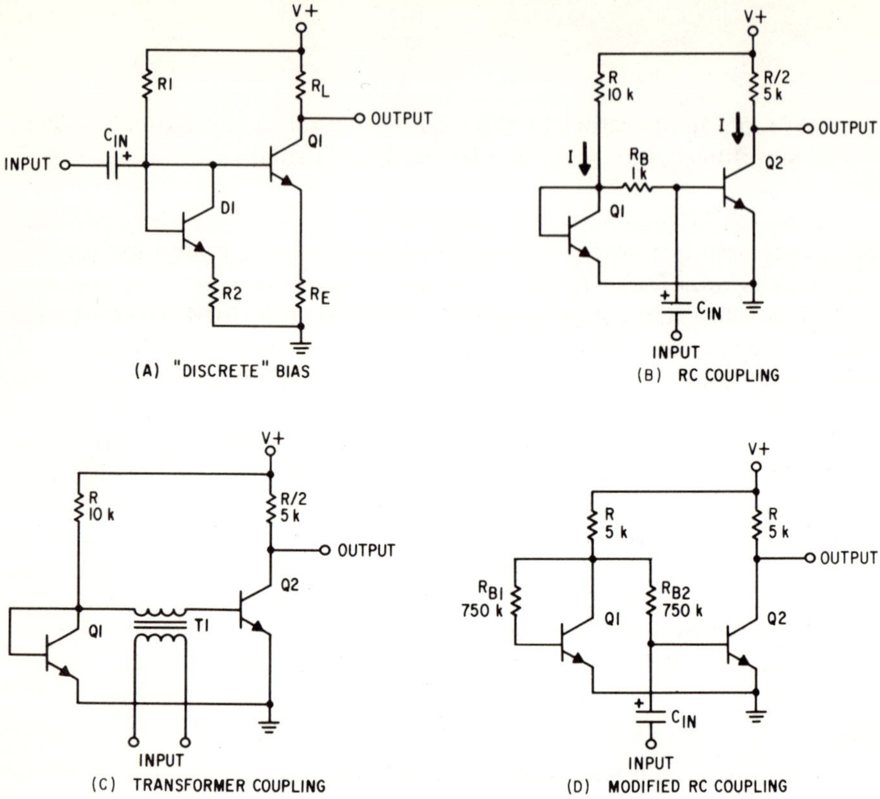

Fig. 2-19 *Biasing techniques: (a) "discrete bias," (b) RC coupling, (c) transformer coupling, and (d) modified RC coupling*

Darlington-connected devices for Q1 and Q2, but these reduce its efficiency of parts.

The transformer-coupled version of Fig. 2-19(c) eliminates the resistor R_B by using the low secondary impedance of coupling transformer T1. This version is well suited to RF or IF applications of array types, for which transformers are small and inexpensive.

A modified R_C coupling scheme that raises input impedance is shown in Fig. 2-19(d). This circuit provides the same desirable bias stability, but without restrictions on input impedance or coupling method. Here, Q1 is forced to operate at a V_{CE} level of $\frac{1}{2}$ V+ by making R_{B1} drop $\frac{1}{2}$ V+, minus one diode drop. A similarly connected resistor, R_{B2} forces Q2 to operate at the same current, since Q1 and Q2 have similar gains. Note that the input impedance in this case can be high.

Differential Amplifier Stage

The differential amplifier (diff amp) is a standard circuit block of linear IC circuitry, even more so than the single-ended stages just discussed, because the differential amplifier lends itself much more readily to dc amplification and the balanced biasing inherent to IC circuitry.

A wide range of circuits are possible using the diff amp as a basis, from simple comparators and op amps to RF amplifiers and limiters.

A basic differential amplifier circuit is shown in Fig. 2-20(a). As shown here, the circuit is used for differential input–differential output operation. This is the most advantageous method, as it takes advantage of the full gain of the circuit and provides maximum noise rejection of undesired signals.

With a regulated source of current, I_E, at the emitter of Q1–Q2, the circuit does not respond to signals common to both input terminals, only to their *difference*. Consequently, the bases of Q1–Q2 can be dc-biased to quiescent voltages over a wide range.

The differential input-to-output gain of the circuit is proportional to R_L and I_E and inversely proportionally to temperature, with the constants K and q, as follows:

$$A_{V_D} = \frac{R_L I_E}{2\left(\frac{KT}{q}\right)}$$

At 300°K, the gain expression can be simplified as follows:

$$A_{V_D} \simeq 19.3\ R_L I_E$$

The gain from input to a *single* output is one-half the differential gain, or

$$A_V = \frac{R_L I_E}{4\left(\frac{KT}{q}\right)}$$

or, at 300°K,

$$A_V \simeq 9.7\ R_L I_E$$

The reader will note from these expressions that for a given current and load resistor, the gain of the differential amplifier is *less* than that of the single-ended amplifier of Fig. 2-18. The difference results from the current split of I_E between input transistors and also from the one-half input-voltage swing split. These reduce the *single-ended* output gain to one-quarter that of the stage of Fig. 2-18. The *differential* mode gain (as shown) is one-half the stage gain in that figure because the effective output swing is doubled between the collectors in this case.

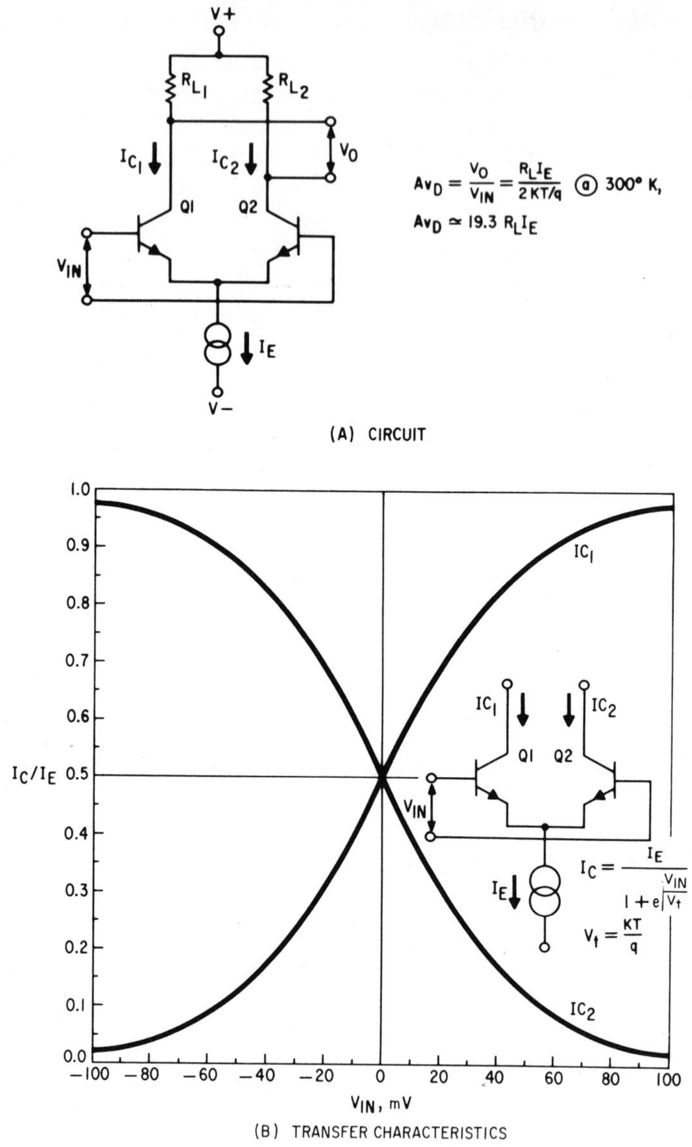

Fig. 2-20 Differential amplifier stage: (a) circuit, and (b) transfer characteristic

Like the single-ended stage previously discussed, the differential stage is linear for small signals only, and gain will be as described. Graphically, the input voltage to current output characteristics of the differential stage can be illustrated, as shown in Fig. 2-20(b).

This figure is a graph of output current I_{C1} (or I_{C2}) versus V_{IN}, where I_C is given by the relation;

$$I_C = \frac{I_E}{1 + e^{\frac{V_{IN}q}{kT}}}$$

At balance, when $V_{IN} = 0$, I_{C1} and I_{C2} are equal at $I_E/2$. As V_{IN} increases, I_{C1} increases while I_{C2} decreases. This situation reverses for decreasing V_{IN}.

As can be noted from the slope of the curve, it is most linear (straightest) in the center region about $V_{IN} = 0$, or for $V_{IN} \simeq \pm 20$ mV. When V_{IN} has reached 100 mV, the current I_{C1} (or I_{C2}) has reached about 98 percent (2 percent) of its final value.

The above tells us several things about the differential amplifier: It is linear for small signals only, and it reaches nearly full swing for inputs of only 100 mV.

A third factor, which can be noted from the fact that the vertical scale factor is *normalized*, is that this basic characteristic holds for *any* operating current I_E (within ratings, of course). However, as the gain expression notes, gain is proportional to I_E; therefore, variation of I_E will result in an overall input–output gain variation. This is a very desirable property of the differential amplifier and does in fact result in a wide variety of useful gain-controlled circuit applications.

A form of the differential amplifier often seen used is the circuit shown in Fig. 2-21. This is simply a standard differential amplifier (Q1–Q2) with a differential to single-ended output converter D1–Q3. D1 and Q3 mirror the current from Q1 and add it (in inverted form) to that from Q2. This subtracts the dc quiescent current and leaves the desired

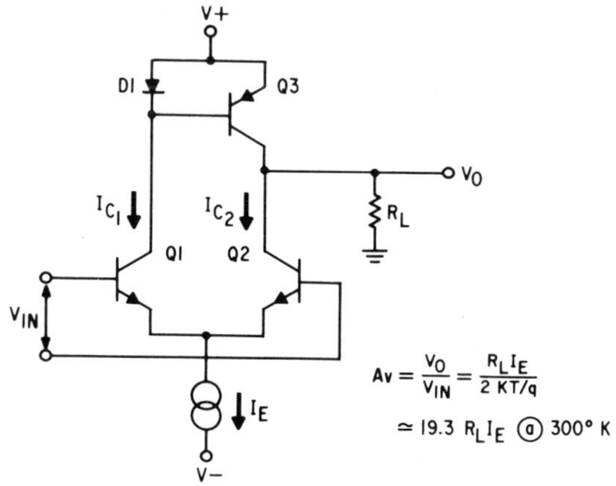

Fig. 2-21 Differential amplifier with differential-to-single-ended output converter

signal current. The gain expression is the same as that for the circuit of Fig. 2-20, but matched load resistors are not required.

Thus far we have discussed only the basic differential amplifier and its transfer/gain characteristics. The signal current output I_0 from a differential amplifier is effectively bipolar ($\pm$) and has a maximum value of I_0 (the emitter current). We can illustrate a more general transfer characteristic as that shown in Fig. 2-22.

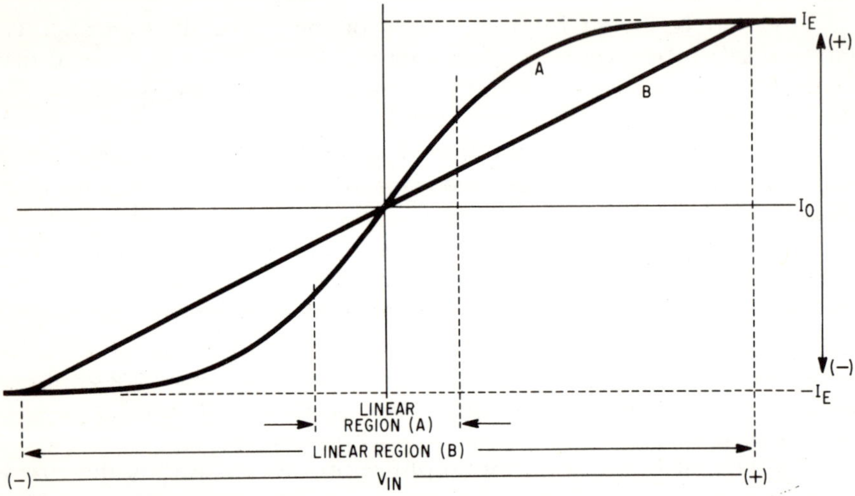

Fig. 2-22 Differential amplifier transfer characteristics for different transconductance

In this sketch, curve A is the output (I_0) that results when we use a differential circuit, as described. Differential amplifiers can also be applied with local feedback, however, as was basically done for the single-ended amplifier. The changes that result are generally like those of the single-ended case—less gain, but more stability and lower distortion.

Curve B of Fig. 2-22 is intended to illustrate how such a reduced gain (lower transconductance) diff amp responds to an input V_{IN}. Note that the maximum output current is still $\pm$ I_E, but the linear operating region is extended to greater input swings because of less gain.

The diff amp with local feedback, as shown in Fig. 2-23, is then desirable when a *linear* transfer characteristic is desired and a large gain is not necessary. The differential output gain expression is given in the figure, including the effect of R_E. As noted previously, when R_E is much greater than the emitter impedance, the gain can be expressed by the simplified equation.

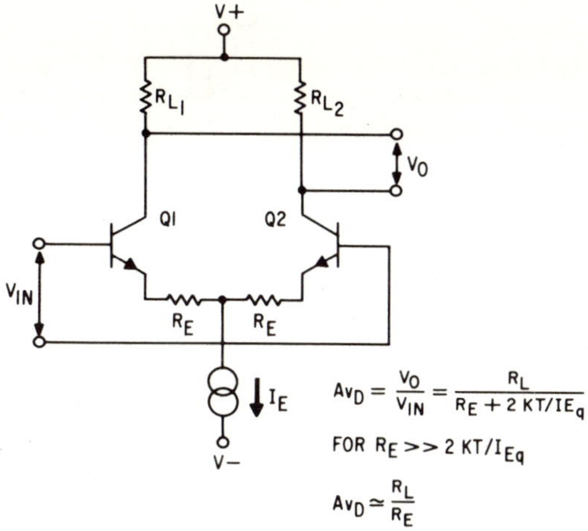

Fig. 2-23 Differential amplifier with local feedback

Biasing Differential Amplifiers

Biasing considerations for a differential amplifier are difficult to generalize upon, since diff amp stages are rarely used alone but rather as parts of a larger circuit. Therefore, input and output biasing will usually be constrained by systems considerations. Examples of such uses are to be found under applications.

A general consideration in applying a diff amp regards the complexity of the current source used. What is required is often dependent upon the nature of the application. A variety of possibilities are illustrated in Fig. 2-24.

The simplest type of current source is a single resistor tied to the V− supply, as in Fig. 2-24(a). This is adequate when the common-mode voltage applied to Q1–Q2 is very low (or zero). An example of this would be when Q1–Q2 is the input stage of an op amp used in the inverting mode. In such a case, the inputs would sit at a fixed potential (usually ground).

For most other diff amp uses, a true current source is required, as shown in the remaining configurations. All of these circuits provide a high impedance to the emitter of Q1–Q2, which causes the common mode rejection to be high. Inasmuch as these current sources have already been discussed, little elaboration is necessary. The equations for I_E are given with the figures.

Two of the most useful of these are the diode-biased [Fig. 2-24(b)] and the Wilson types [Fig. 2-24(c)]. The diode-biased current source can be used in its basic mode (shown) or as a logarithmic type by the addition of R_E (shown dotted). The Wilson type current source is a good choice when the

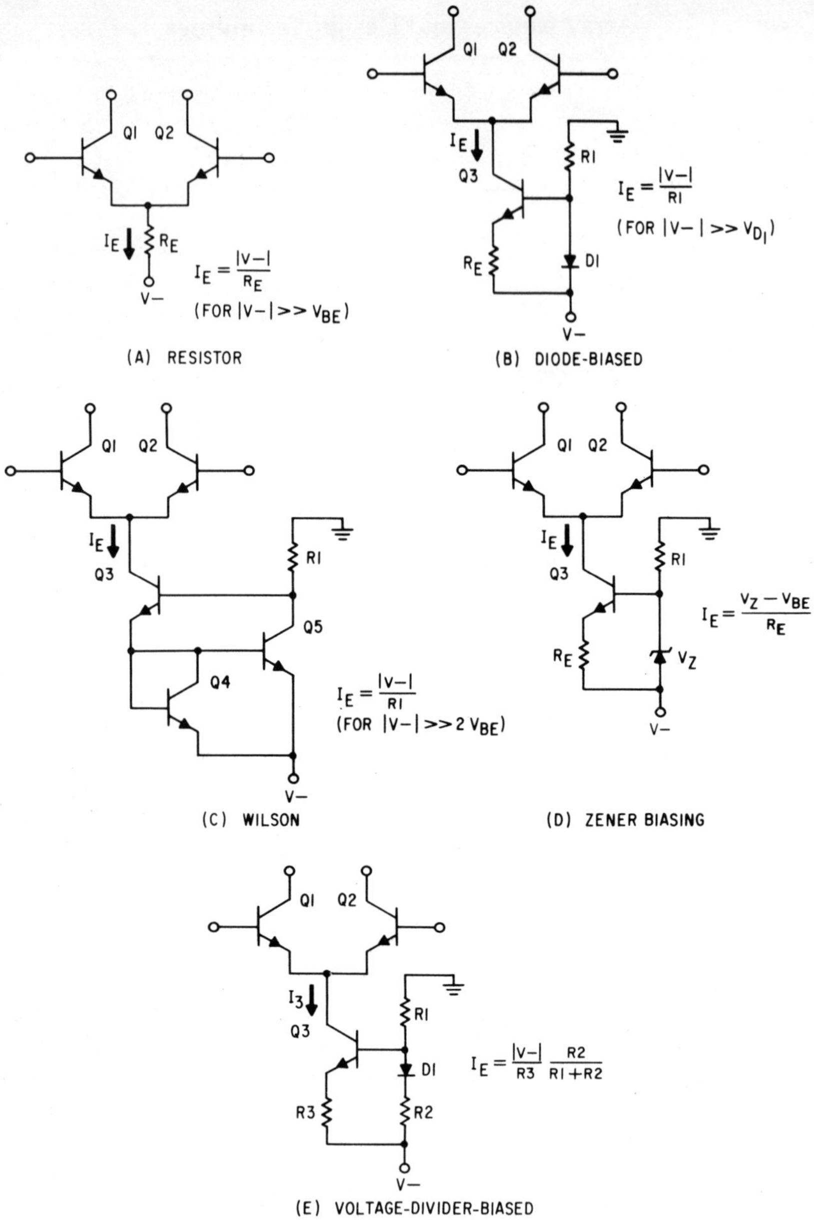

Fig. 2-24 *Current source variations used with differential amplifiers: (a) resistor, (b) diode-biased, (c) Wilson, (d) Zener-biased, and (e) voltage-divider-biased*

highest CMRR is required because of its high output impedance. The remaining two circuits [Figs. 2-24(d) and (e)] offer a reduced number of parts as an advantage but require larger voltages for proper biasing, where types 2-24(b) and 2-24(c) do not.

A consideration arising when the highest performance is required of a diff amp is its input offset voltage. Although array transistors typically exhibit offsets of 1 mV or less, it may be desirable in some cases to null the offset to zero (exactly). A method of accomplishing this is shown in Fig. 2-25.

In this circuit, Q1–Q2 are a diff amp pair that operate at currents of $I_E/2$ at balance. To compensate for up to several mV of offset between Q1–Q2, a pot of R Ω is added between their emitters as shown, with I_E applied to the wiper.

If the pot value is selected as shown, offsets of up to 10 mV can be effectively canceled. Some gain reduction will be introduced by this pot, but it is minimized for the values selected, as shown.

Alternative methods of applying offset connection to a diff amp is to introduce a small (≃ 10 mV) correction voltage to the base of Q1 and Q2 or to unbalance the collector load resistors by a small percentage.

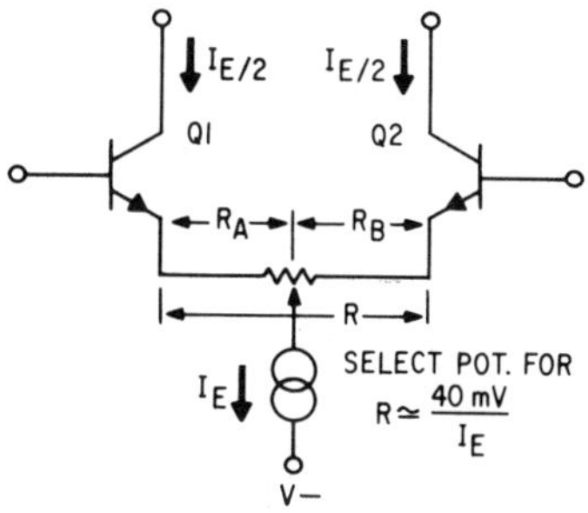

Fig. 2-25 Offset trim technique for differential amplifiers

Composite Stages

Often, the limitations of an IC transistor may prevent its use in a certain mode of operation. Such a situation does not necessarily defeat the use of IC array types, however, as they can readily be operated as *composite stages* that circumvent such obstacles. A composite stage is one where two or more devices are combined in such a manner as to yield the equivalent of a single high-performance unit. A composite stage can be used either alone or within a larger circuit to enhance overall performance. Several composite stage types are illustrated in Fig. 2-26.

In Fig. 2-26(a) are illustrated Darlington-paired devices that are most likely familiar from discrete designs. Darlington connections are employed to increase the net current gain beyond that of a single device and are most useful to pnp stages where typical pnp H_{FE} is usually low. Typical Darlington stage gains are 10,000 or more for npn's and up to 1000 for pnp's. An optional bias resistor (R) is sometimes used to define

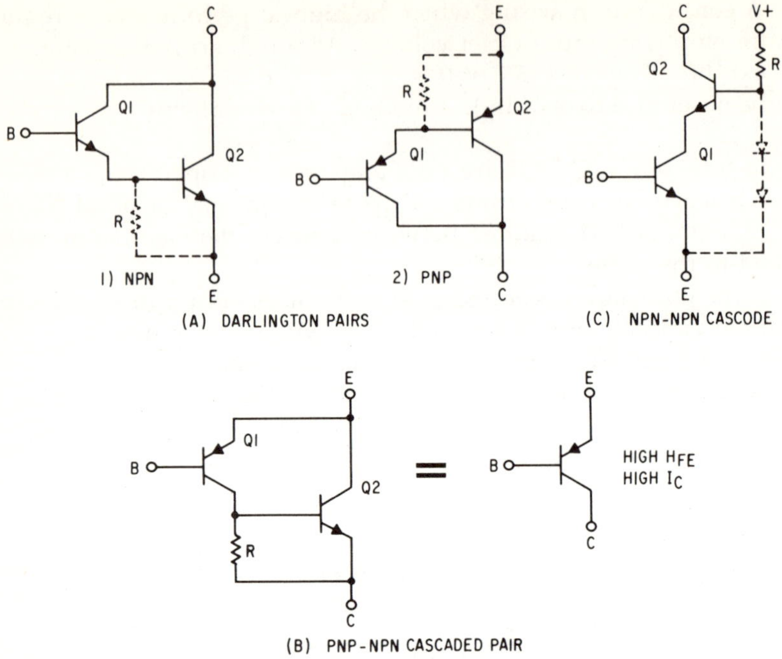

Fig. 2-26 *Composite stages: (a) Darlington pairs, (b) npn-npn cascode, and (c) pnp-npn cascaded pair*

the emitter current of Q1 or to control leakage of Q2 at high temperatures.

A very useful connection is the pnp–npn cascade of Fig. 2-26(b). This circuit produces the functional equivalent of a medium-power pnp transistor, a device nonexistent in array ICs. Q2 is a medium-power npn, and Q1 is a small signal pnp. The combination has an H_{FE} that is the product of the individual gains and a collector rating that is equal to that of Q2. This circuit is often used in class AB or B output stages for load drive ability. Bias resistor R is selected to control the emitter current of Q1.

The npn-npn cascade of Fig. 2-26(c) is used in wide-band stages where parasitic capacity must be minimized. Q1 operates as a common emitter stage, driving Q2 as a common base stage. The combination is equivalent to a wide-bandwidth npn device, with no collector-to-base feedback capacitance around Q1. Base bias is necessary for Q1 and can be similar to that shown.

An arrangement similar to this is also used in "super-beta" array types (such as the CA3095). Here, the primary purpose of the cascade is to lower the CE voltage of Q1, as super-beta transistors have very low V_{CE} ratings. In the circuit of Fig. 2-26(c), the V_{CE} rating of the composite is that of Q2, whereas the gain is that of Q1.

Buffer Stages

There are various types of buffer stages used with IC array circuits, but all have one thing in common, that is, to preserve the fidelity of a signal while driving a practical (usually low) load impedance. Most buffer stages have unity voltage gain but may have *current* gains of several thousand or more. Load currents practical with array transistor types range up to 1 A, with voltages up to 40 V.

Single-ended buffers

The most straightforward buffer application is the standard class A, single-ended emitter follower. A number of variations of this circuit are shown in Fig. 2-27.

In Fig. 2-27(a) is shown a simple npn emitter follower, with a resistor to establish the quiescent current. This circuit is adequate when the putput swing is small or the load current is relatively light.

For high-level output swing, the circuit of Fig. 2-27(b) is more appropriate, as it adds a constant current bias, I_E. This bias current can be set to just over the maximum required for negative signal peaks. Since I_E will not increase for positive swings [as will the current in R_E of Fig. 2-27(a)], the power dissipation of this stage can be lower, and distortion also because of less signal current variation in Q1.

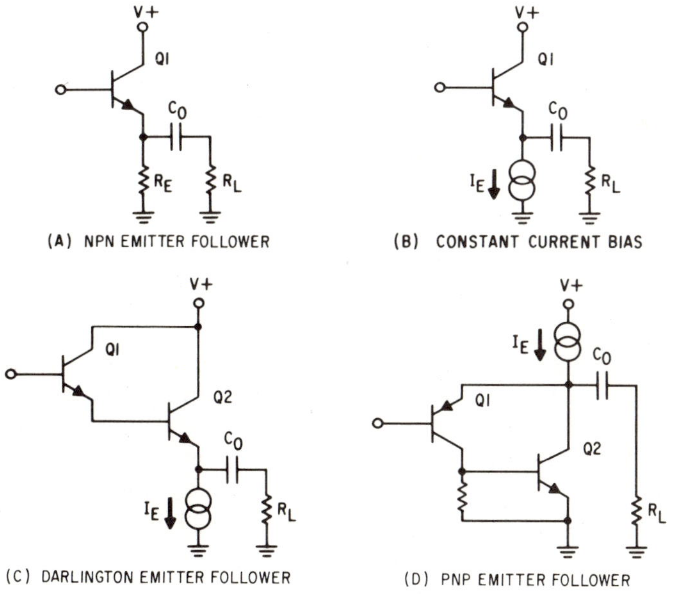

Fig. 2-27 Class A single-ended buffer stages: (a) npn emitter follower, (b) constant current bias, (c) Darlington emitter follower, and (d) npn emitter follower

The Darlington-connected emitter follower of Fig. 2-27(c) improves the circuit further by increasing the current gain of the stage and further isolating loading from the source-driving Q1.

All three of these three stages can be wide bandwidth, with response up to several MHz. The best frequency response will be attained for emitter currents of several mA, where transistor f_T is maximum.

The composite pnp emitter follower of Fig. 2-27(d) is not capable of wide bandwidth because of the use of a pnp Q1. Its advantage is in better linearity resulting from the 100-percent feedback to the emitter of Q1. Distortion of this stage can be very low—much better than 0.1 percent.

For driving heavy currents into low impedance loads with minimal standby power, a class AB output stage is preferred. Two variations on this type of stage are shown in Fig. 2-28.

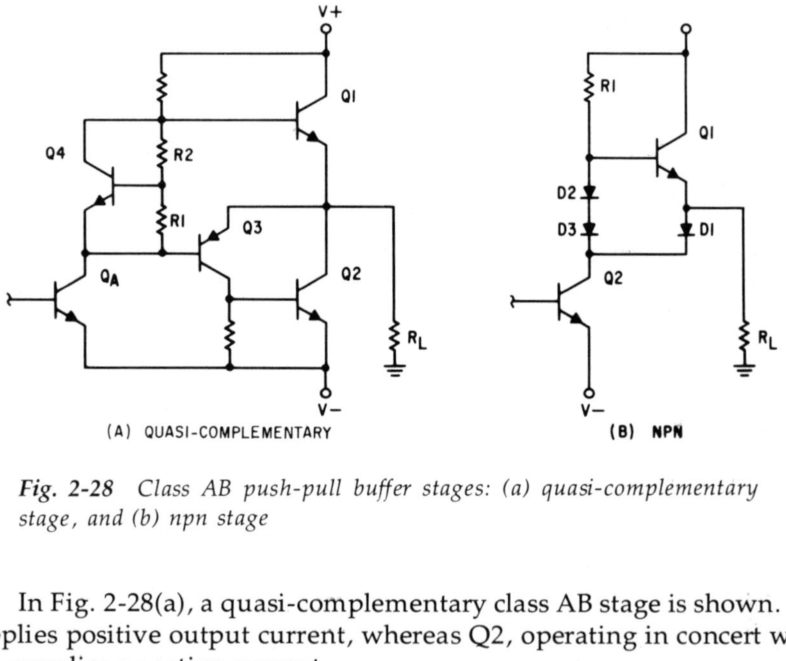

Fig. 2-28 *Class AB push-pull buffer stages: (a) quasi-complementary stage, and (b) npn stage*

In Fig. 2-28(a), a quasi-complementary class AB stage is shown. Q1 supplies positive output current, whereas Q2, operating in concert with Q3, supplies negative current.

Class AB stages are usually biased just on the verge of common-mode conduction (from V+ to V−) to minimize standby power. In this circuit, the requisite base bias for Q1 and Q3–Q4 is established by Q4, a V_{BE} multiplier. When R2 = R1, 2 V_{BE} is applied between Q1 and Q3, causing Q1 and Q3–Q2 to idle at a current equal to that in QA. With Q4, Q1, and Q3 monolithically matched, this current will track with temperature.

The all npn stage shown in Fig. 2-28(b) has the advantage of wider bandwidth because of the elimination of the pnp from the signal path. In this circuit, Q2 provides negative output current through D1, and D2–D3 provides a temperature compensated bias voltage for Q1. If Q1 and D1–D3 are on the same chip, the idle current will be stable.

Over-Current Protection

It is usually prudent to provide safety precautions in circuits using ICs to prevent the inevitable accidents that can destroy them. The most frequent cause of failure is over current, which fuses internal IC connections. Fortunately, simple precautionary measures exist to prevent such accidents from inflicting permanent damage, as shown in Fig. 2-29.

For small signal transistors operating in low power stages, the simple protection scheme of Fig. 2-29(a) will probably be satisfactory. Here a resistor is added to the collector of the emitter-follower output stage, which limits the short-circuit current to a safe level. In some cases, the resistor might be included in series with the output lead, but this may be a disadvantage, as it will produce a drop in signal level.

A push-pull stage can be protected as shown in Fig. 2-29(b). Here, protection resistors R_S (+) and R_S (−) are inserted in the collector leads of the (+) and (−) output devices, Q1 and Q2. An alternative form of protection is a single resistor in series with the output lead.

Simple series-resistor protection schemes such as these will cause the current limit level to vary with load and supply voltage, since they basically set up a voltage divider action. Current limiting that is independent of voltage or load can be obtained by using the circuit of Fig. 2-29(c). Here, sample resistors R1 and R2 monitor the (+) and (−) output currents; in the case of over current, they trip either Q3 or Q4 on and thus Q1 and Q2 either off or to a safe current level.

Parasitic Action

One of the application pitfalls encountered in using IC array transistors that is not encountered with discretes is undesired parasitics. Parasitic action can take place as a functional (but undesired) diode or transistor in addition to the desired function. It is due simply to the junction isolation process used to build this type of IC. Potential parasitic action cannot be ignored if one aims for predictable, reliable behavior.

A common example of how undesired parasitic pnp action can occur in conjunction with normal npn behavior is seen in Fig. 2-30(a), which shows the equivalent schematic diagram of the desired npn and the undesired (parasitic) pnp. A cross-sectional view is shown in Fig. 2-30(b), which illustrates how this occurs.

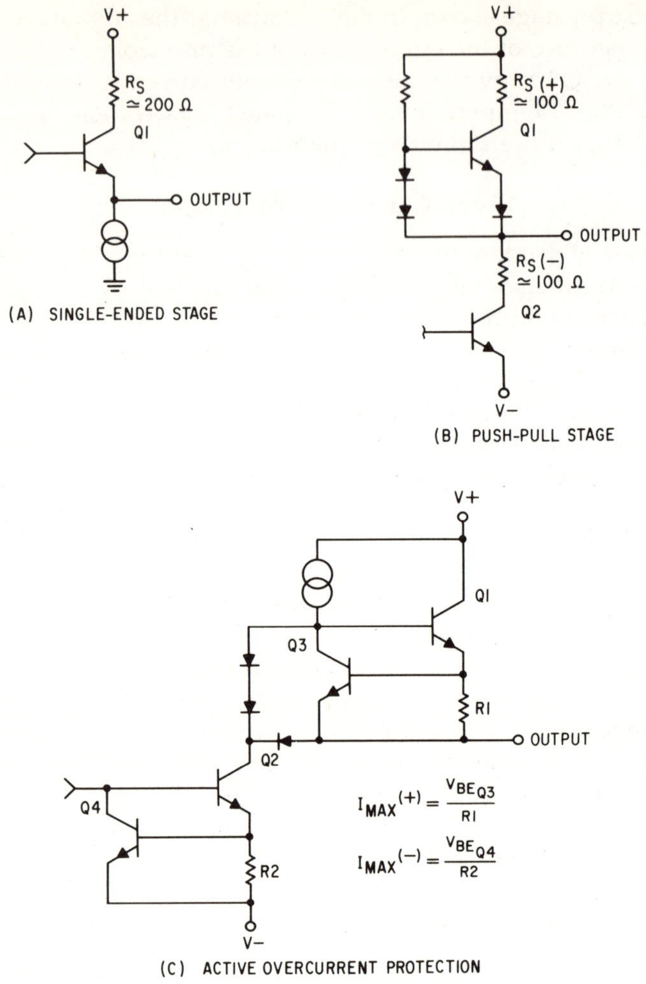

Fig. 2-29 *Overcurrent protection: (a) single-ended stage, (b) push-pull stage, and (c) active overcurrent protection*

From Fig. 2-30(b), it can be noted that in the isolation well formed for the npn there is not only the (desired) npn transistor junctions between E, B, and C, but also a pnp between the npn base, npn collector, and the substrate, as circled. Unless the biasing conditions of the npn are controlled carefully, this pnp can be activated, thus disrupting circuit behavior.

As the schematic and cross-sectional view show, this parasitic pnp is unavoidable with this type of process. Not only does a so-called "substrate pnp" exist as just described, but there is also a *lateral* parasi-

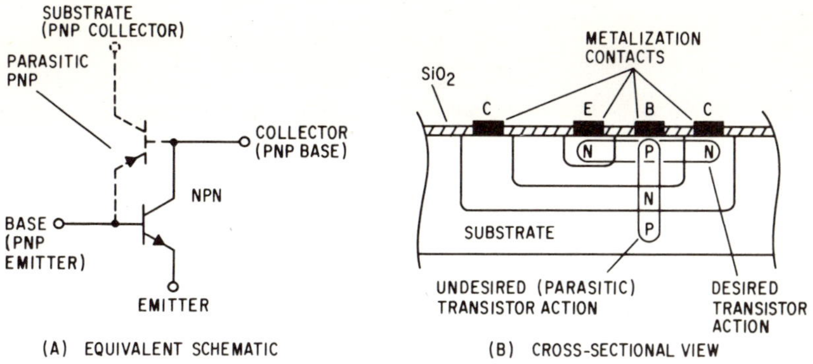

Fig. 2-30 *Parasitic action in monolithic IC transistors: (a) equivalent schematic, and (b) cross section*

tic pnp existing to the side wall of the isolation well (which is common to the substrate). Circuit-wise conduction of this pnp can be controlled by avoiding base-collector conduction of the npn—which, as can be noted, forms the pnp emitter-base junction—and also by limiting the degree of current available to the npn. A factor that aids in the control of the pnp's conduction is its relatively low H_{FE} (usually in the range of 5 to 10).

An even more common potential parasitic is the collector-to-substrate diode inherent in every npn transistor. In Fig. 2-30(a), this diode is represented by the base-collector junction of the pnp. Diode action can exist at this junction, of course—whether or not the npn is conducting—if the substrate is allowed to be more positive than the npn collector. The rule for avoiding this most basic parasitic is to guarantee that the substrate is *always* at the most negative of the applied circuit potential. Since the substrate is common to all transistors in a given IC, violation of this bias condition can disrupt the proper function of any (or all) transistors of the IC. Note that the "most negative bias level" rule for the substrate applies on a transient as well as steady-state (dc) basis.

Signal and Zener diodes using npn transistors are subject to parasitic action also if the connections are not properly controlled. This case is illustrated in Fig. 2-31.

The preferred signal diode or rectifier connection and the connection to avoid are shown in Fig. 2-31(a). Shorting the base-collector junction of an npn gives a stable predictable diode with well-defined characteristics. The open-collector base-emitter connection should be avoided, as the floating collector is subject to leakage and undesired parasitics.

Similarly, the preferred Zener connection is shown in Fig. 2-31(b), with the collector and emitter shorted. The connection that should definitely be avoided will (or may) incur accompanying parasitic conduction.

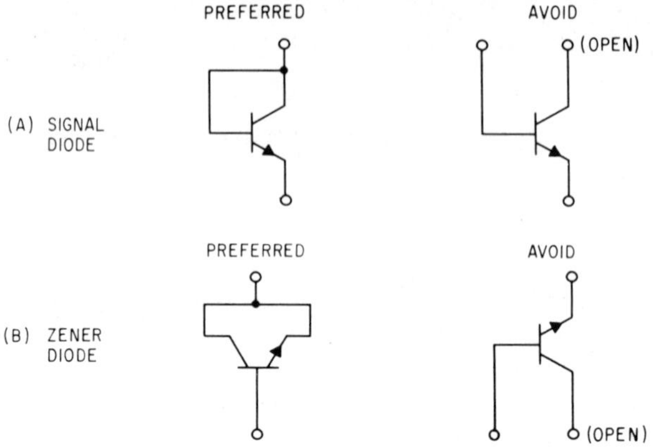

Fig. 2-31 *Connection of array devices to avoid parasitics: (a) signal diode, and (b) Zener diode*

A good summary comment on parasitics in IC arrays is that they are *always* a potential hazard and should be considered possible in every circuit regardless of its simplicity.

REFERENCES FOR CHAPTER 2

(1) Widlar, R. J., "Some Circuit Design Techniques for Linear Integrated Circuits," *IEEE Transactions on Circuit Theory*, Dec. 1965.

(2) Kleinman, H. M., "Design Ideas for RCA Linear Arrays," RCA Application Note ST-3895, Dec. 1968.

(3) Hoover, M. V., "Monolithic Linear IC Arrays and Their Applications," RCA Application Note ST-3958, March 1969.

(4) Camenzind, H. R., and Grebene, A. B., "An Outline of Design Techniques for Linear Integrated Circuits," *IEEE JSSC*, June 1969.

(5) Callahan, M. J., "Charts Speed the Designing of Constant Current Sources," *Electronics*, Aug. 17, 1970.

(6) Sullivan D. R., and Maidique, M. A., "Matched Design-Key to Linear ICs," *EDN*, Feb. 15, 1971.

(7) Wilson, G. R., "A Monolithic Junction FET-NPN Operational Amplifier" *IEEE ISSCC Digest*, Feb. 1968.

(8) Wittlinger, H. A., "Applications of the CA3080 and CA3080A High Performance Transconductance Amplifiers," RCA ICAN-6668, Sept. 1971.

(9) Rieger, M. L., "Solve DC Current Source Equations," *Electronic Design*, Oct. 25, 1977.

(10) Allen, P. E., "Graphical Analysis of Matched Transistor Current Sinks/Sources," *IEEE JSSC*, Feb. 1974.

(11) Widlar, R. J., "A Versatile Monolithic Voltage Regulator," National Semiconductor AN-1, 1967.

(12) Lieux, J. D., and Ricks, R. D., "IC Regulator Removes Restrictions," *The Electronic Engineer*, March 1969.

(13) Widlar, R. J., "IC Provides On-Card Regulator for Logic Circuits," National Semiconductor AN42, Feb. 1971.

(14) Dobkin, R. C., "1.2 Volt Reference," National Semiconductor AN-56, Dec. 1971.

(15) Brokaw, A. P., "A Simple Three Terminal IC Bandgap Reference," *IEEE JSSC*, Dec. 1974.

(16) Kuijk, K. E., "A Precision Voltage Reference Source," *IEEE JSSC*, June 1973.

(17) ______, "Low Cost Voltage Reference for Analog Circuits," *Analog Dialogue*, 8-2, 1974.

(18) Keller, J. P. , "Differential Amplifiers at Work," *Electronics*, Sept. 18, 1967.

(19) Lancaster, D. E., "The Differential Amplifier," *Electronics World*, Feb. 1968.

(20) ______, "Linear Integrated Circuits," RCA Technical Series IC-42, 1970.

(21) Grebene, A. B., *Analog Integrated Circuit Design*, Van Nostrand Reinhold, 1972.

(22) Camenzind, H. R., *Electronic Integrated Systems Design*" Van Nostrand Reinhold, 1972.

(23) ______, "101 Analog IC Designs," *Interdesign*, 1976.

3
IC Array Mini-Catalog

This chapter briefly discusses the types of IC array devices featured by the applications section of Part II. All these devices are representative, readily available units, usually available from multiple sources. The information is in the form of basic package pinouts and diagrams that depict the functional arrangement of each part. Capsule specifications are also given.

To supplement this information, data sheets for selected devices are included in the Appendix along with a manufacturers' source listing for all devices.

General-Purpose Arrays

There are a number of array devices that fall into the category of general-purpose units. As might be expected, these units are the most popular.

The two most popular array types are the CA3046 and CA3018, shown in Figs. 3-1 and 3-2, respectively. The 3046 is a five-transistor, 14-pin (DIP) array. The 3018 is a four-transistor, 12-pin (TO-5) unit. Both types have derivatives: The CA3086 is a more loosely spec'd 3046, whereas the CA3146 and CA3118 are 30-V max versions of their counterparts. The CA3146A and CA3118A increase the voltage rating to 40 V (max).

Current ratings for these devices is listed as 50 mA (max), but operation should be limited to 10 mA or less for most predictable operation. Data sheets for these devices are included in the Appendix.

The CA3084 device (Fig. 3-3) is an all pnp array, using standard lateral pnp transistors. As can be noted, there are six devices, with two uncommitted and the remaining four in Darlington and current mirror hookups. Gain of the individual transistors is low, as is typical of lateral pnp's; voltage ratings are 40 V (min).

The CA3096 array (Fig. 3-4) features (uncommitted) three npn and two pnp transistors. All the devices have quite respectable gains and voltage ratings, as can be noted. The CA3096A is a tighter spec'd version of the basic device. The CA3096 and CA3096A are supplied in 16-pin DIPs. A CA3096 data sheet is included in the Appendix.

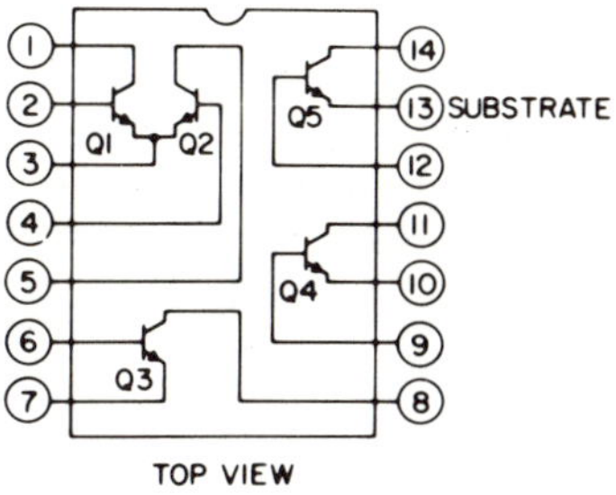

	$V_{(BR)CEO}$	$V_{(BR)CBO}$	h_{FE}	I_C	$\lvert V_{IO} \rvert$
CA3045/46	15V Min.	20V Min.	40 Min.	50mA Max.	0.45mV (TYP.)
CA3086	15V Min.	20V Min.	40 Min.	50mA Max.	0.45mV (TYP.)
CA3146	30V Min.	40V Min.	30 Min.	50mA Max.	0.48mV (TYP.)
CA3146A	40V Min.	50V Min.	30 Min.	50mA Max.	0.48mV (TYP.)

Fig. 3-1 CA3046, CA3086, CA3045, and CA3146 (3146A) array devices

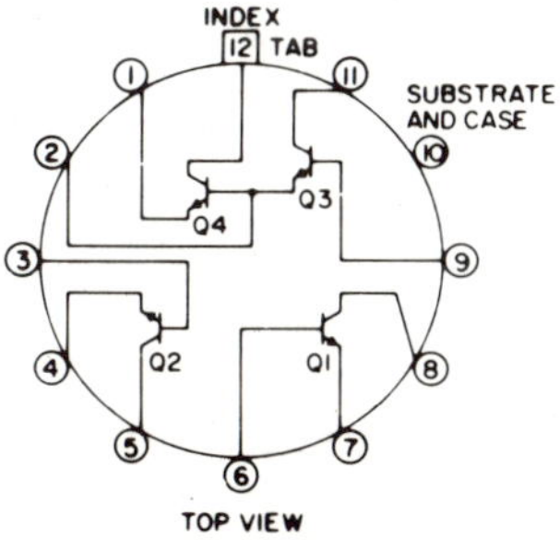

	$V_{(BR)CEO}$	$V_{(BR)CBO}$	h_{FE}	I_C
CA3018	15V Max.	20V Max.	30 Min.	50mA Max.
CA3018A	15V Max.	30V Max.	60 Min.	50mA Max.
CA3118	30V Max.	40V Max.	30 Min.	50mA Max.
CA3118A	40V Max.	50V Max.	30 Min.	50mA Max.

Fig. 3-2 CA3018 (3018A) and CA3118 (3118A) array devices

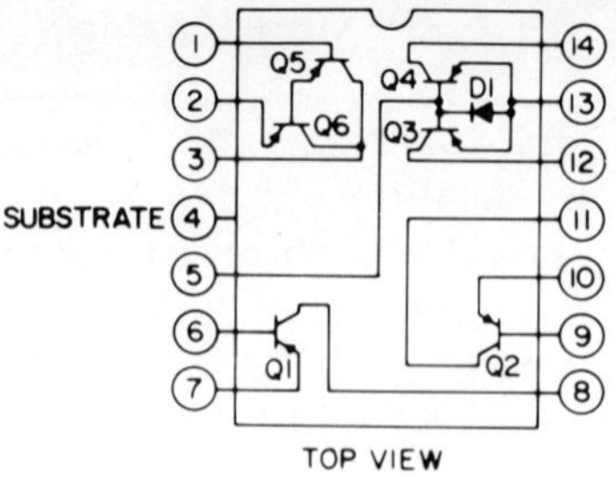

$V_{(BR)CEO}$ = −40V Min.
$V_{(BR)CBO}$ = −40V Min.
h_{FE} = 15 Min.
I_C = −10 mA Max.

Fig. 3-3 *CA3084 array device*

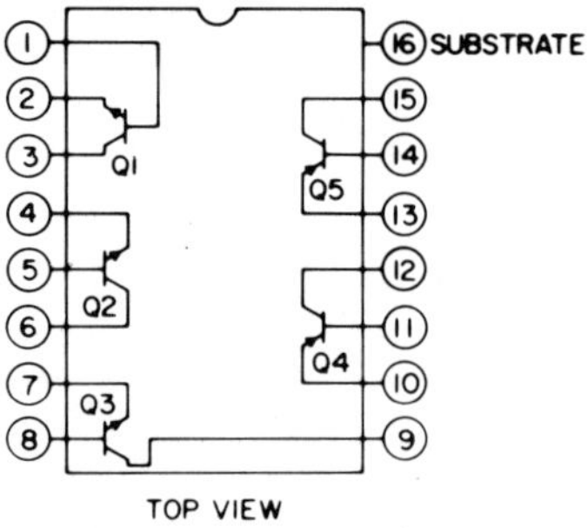

	$V_{(BR)CEO}$	$V_{(BR)CBO}$	h_{FE}	I_C
Each n-p-n	35V Min.	45V Min.	150 Min.	50mA Max.
Each p-n-p	−40V Min.	−40V Min.	40 Min.	−10mA Max.

NOTE: The leakage limits on the CA3096A are tighter than on the CA3096.

Fig. 3-4 *CA3096 (3096A) array device*

High-Current Devices

There are a number of high-current array types. The CA3083 shown in Fig. 3-5 is an uncommitted five-transistor npn array, with a current rating of 100 mA and a voltage rating of 15 V. The CA3183 increases the voltage rating to 30 V but has a current limit of 75 mA. The 3183A is similar, but with a 40-V rating. All these types are supplied in a 16-pin DIP.

The CA3081 and CA3082 in Fig. 3-6 trade the flexibility of the 3083 for a greater transistor density. The CA3081 (left) features seven transistors in a common-emitter arrangement. The CA3082 (right) features seven transistors also, but connected in a common-collector

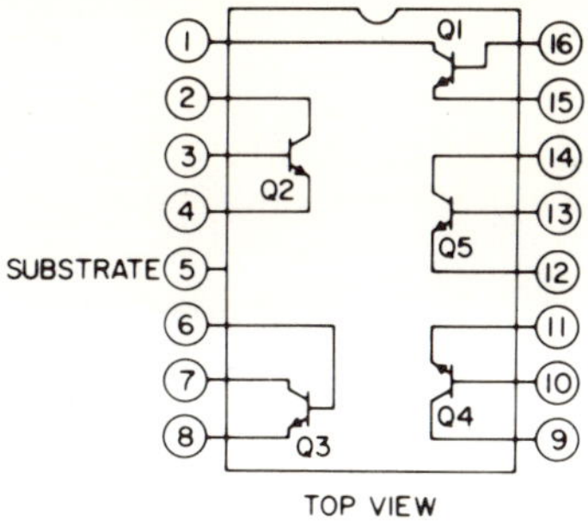

	$V_{(BR)CEO}$	$V_{(BR)CBO}$	h_{FE}	I_C
CA3083	15V Min.	20V Min.	40 Min.	100mA Max.
CA3183	30V Min.	40V Min.	40 Min.	75mA Max.
CA3183A	40V Min.	50V Min.	40 Min.	75mA Max.

Fig. 3-5 *CA3083 and CA3183 (3183A) array devices*

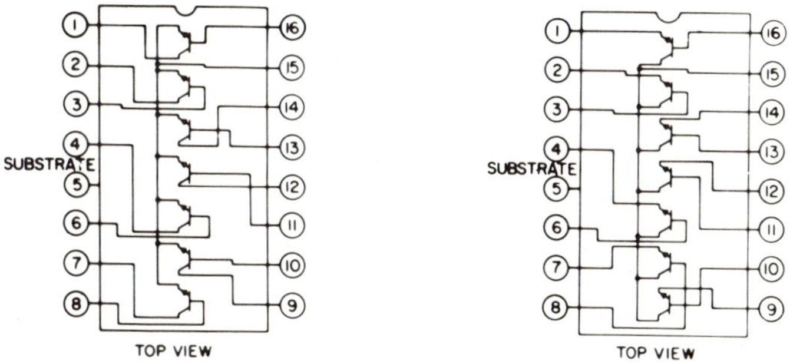

$V_{(BR)CEO}$ = 16V Min.
$V_{(BR)CES}$ = 20V Min.
h_{FE} = 40 Min.
I_C = 100 mA Max.

Fig. 3-6 *(a)CA3081, common emitter; and (b) CA3082, common collector*

array. Both types have 16-V, 100-mA ratings and are contained in 16-pin DIPs.

The CA3724, CA3725, and CA3138 arrays shown in Fig. 3-7 feature four npn transistors with 1-A current ratings. The voltage ratings differ—being 50, 40, and 15 V (min), respectively. All types are contained in 14-pin DIPs. These devices, it should be noted, are *not* monolithically matched (as are the others).

Mixed Element

There are several IC array types that offer a mixture of elements such as diodes, Zener diodes, npn and pnp transistors, SCRs, and even PUTs (programmable unjunction transistors).

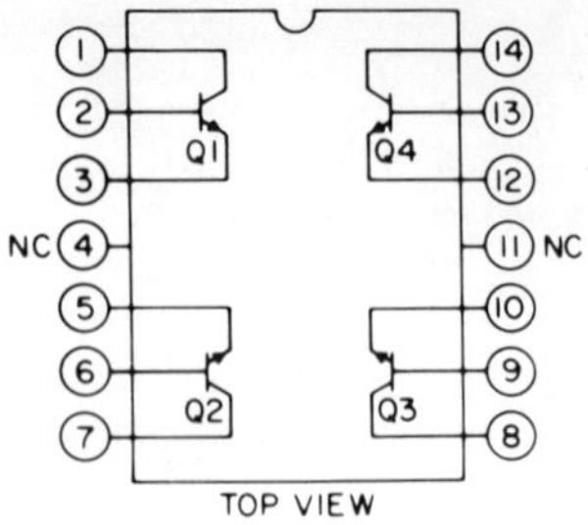

	$V_{(BR)CEO}$	$V_{(BR)CBO}$	h_{FE}	I_C
CA3724G	40V Min.	70V Min.	35 Min.	1.0 A Max.
CA3725G	50V Min.	80V Min.	35 Min.	1.0 A Max.
CA3138G	15V Min.	20V Min.	70 Min.	1.0 A Max.

Fig. 3-7 *CA3724, CA3725, and CA3138 devices*

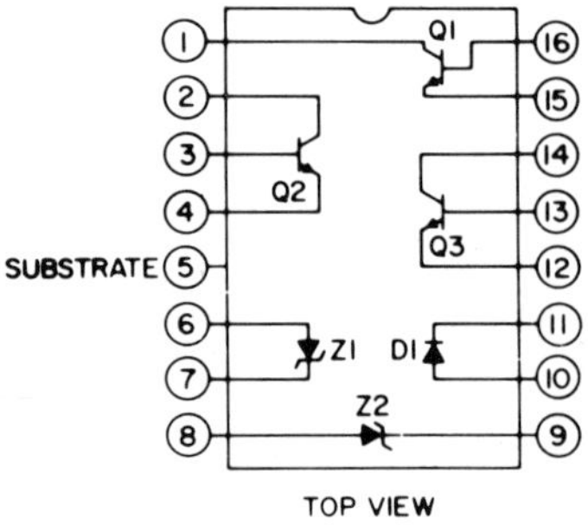

Transistors
$V_{(BR)CEO}$ = 15V Min.
$V_{(BR)CBO}$ = 20V Min.
h_{FE} = 40 Min.
I_C = 100mA Max

Zener Diodes
V_z = 7V ± 10%

Diode
$V_{(BR)DR}$ = 5.5V Min.

Fig. 3-8 *CA3093 array device*

The CA3093, shown in Fig. 3-8, has three 100-mA current rating npn transistors, two Zeners, and a single forward diode. The transistors are similar to those of the CA3083; the Zeners have a 7 V ± 10 percent voltage. The CA3093 is contained in a 16-pin DIP.

The CA3097 is a veritable smorgasbord of devices, as shown in Fig. 3-9. This device has a single npn, an SCR, a PUT, and an npn-pnp composite pair. Voltages ratages ratings are 30 V (npn) and 40 V (pnp). The device is packaged in a 16-pin DIP.

Super-Beta Array

The CA3095 is illustrated in Fig. 3-10. This device features a differentially connected pair of super-beta transistors (Q1 and Q2); three medium-voltage, standard npn types; and a built-in cascade-biasing network for the super-beta units. The super-beta units feature a minimum gain of 1,000; the standard devices have gains of 100. The CA3095 is contained in a 16-pin DIP.

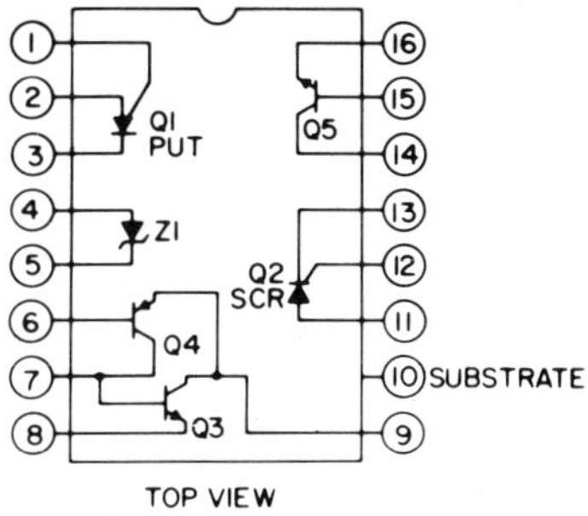

$V_{(BR)CEO}$ = 30V Min. (n-p-n)
= −40V Min. (p-n-p)
$V_{(BR)CBO}$ = 50V Min. (n-p-n)
= −50V Min. (p-n-p)
I_C = 100mA Min. (n-p-n)
= −10mA Min. (p-n-p)

p-n-p/n-p-n pair: h_{FE} = 8000 (TYP.)
PUT: I_p = 15nA (TYP.), V_{AK} = ± 30V Max.
SCR: $I_{T(DC)}$ = 150mA Max.
V_{DRXM} = 30V
Zener: 8V ± 10%

Fig. 3-9 CA3097 array device

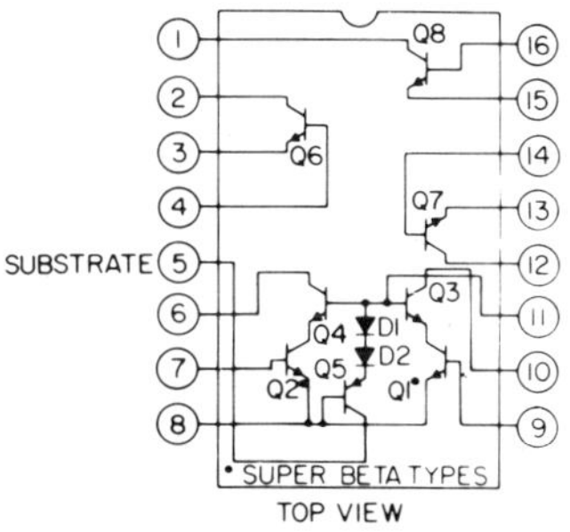

For conventional transistors:
$V_{(BR)CEO}$ = 35V Min.
$V_{(BR)CBO}$ = 45V Min.
h_{FE} = 150 Min.
I_C = 50mA Max.
For super-beta transitor pairs:
h_{FE} = 1000 Min.

Fig. 3-10 CA3095 array device

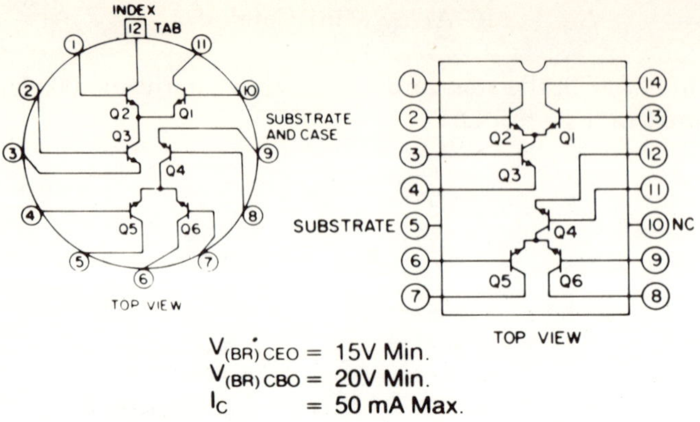

Fig. 3-11 *CA3026 and CA3054 array devices*

Differential Amplifier Arrays

There are several dual differential npn amplifier arrays available that offer either standard-process devices with 500-MHz f_T's or extended-frequency devices with f_T's of 1 GHz.

The CA3026 and CA3054 devices are shown in Fig. 3-11. They are constructed with general-purpose npn devices, similar to those within 3046 devices. The CA3026 is contained in a 12-pin TO-5; the CA3054, in a 14-pin DIP.

The CA3049 and CA3102 are shown in Fig. 3-12. They feature an f_T capability in excess of 1 GHz. Voltage and current ratings are otherwise similar to those of the general-purpose units. Note that the CA3049 is a pin equivalent to the CA3026 but that the CA3102 pinout is different from that of the CA3054 because of layout screening considerations.

High-Frequency Array

The CA3127 array, shown in Fig. 3-13, is an uncommitted array of five npn transistors. These devices have an extended f_T capability, similar to those of the CA3049 and CA3102 units. The device is supplied in a 16-pin DIP.

Diode Arrays

There are a variety of diode arrays available, with options as to pinout and functional arrangement, breakdown voltage, and reverse leakage.

The CA3039, shown in Fig. 3-14, features six uncommitted diodes, one to substrate. These are diode-connected transistors; therefore, the reverse voltage applied must be less than V_{EB}. Leakage is controlled to 100 nA (max) at a 5-V reverse voltage. Matching is within 5 mV. The CA3039 is supplied in a 12-pin TO-5 package.

The CA3019 is a quad-plus-two single diode array, as shown in

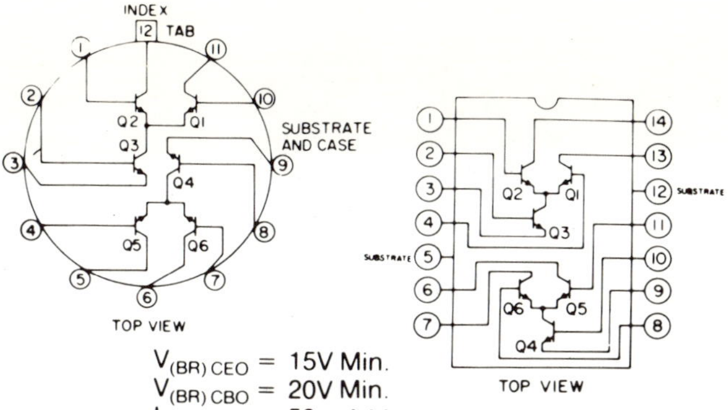

$V_{(BR)CEO}$ = 15V Min.
$V_{(BR)CBO}$ = 20V Min.
I_C = 50 mA Max.
Power Gain = 23 dB TYP. at 200 MH_z

Fig. 3-12 CA3049 and CA3102 array devices

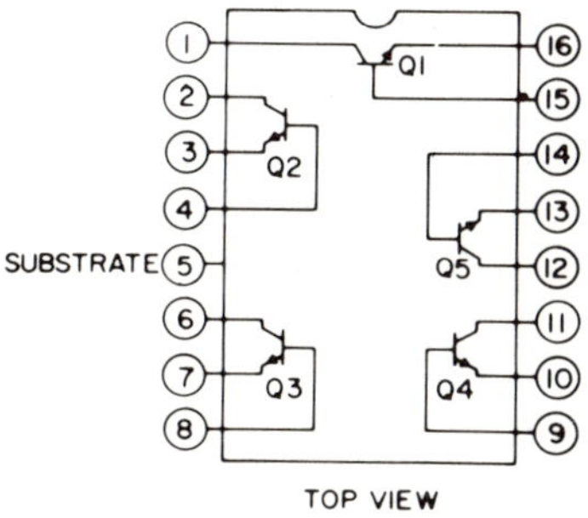

$V_{(BR)CEO.}$ = 15V Min.
$V_{(BR)CBO}$ = 20V Min.
h_{FE} = 40 Min.
I_C = 20mA Max.
f_T = 1.15 GH_z (TYP.)
NF = 3.5 dB (TYP.) at 100 MH_z

Fig. 3-13 CA3127 array device

Fig. 3-15. These are also diode-connected transistors, and the reverse voltage is 4-V max. The allowable leakage of the CA3019 is higher than the CA3039—10 μA (max). Matching is similar at 5 mV, and the CA3019 is also supplied in a 12-pin TO-5 package.

The CA3141 shown in Fig. 3-16 is a higher voltage array of 10 diodes, connected as shown. These devices are transistor collector-base junctions; therefore, there is a small loss of diode output (cathode) current because of parasitic substrate conduction. This loss is typically 4 percent as specified by the 96-percent current match. The CA3141 is supplied in a 16-pin DIP.

Technical data for this chapter was supplied through the courtesy of RCA Solid State Division.

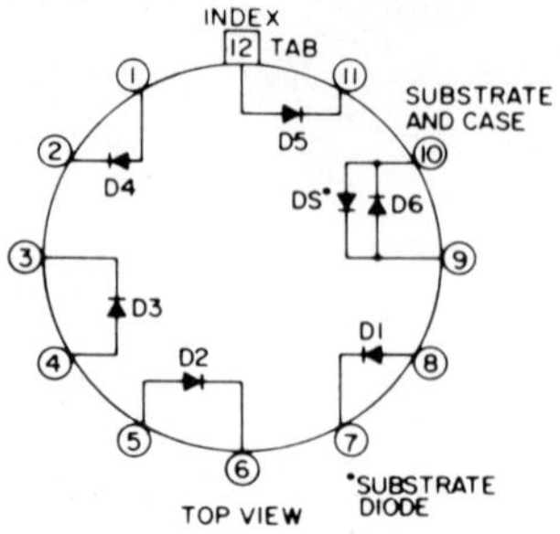

$V_{(BR)R}$ = 5V Min.
I_R = 100 nA Max.
$|V_{F1} - V_{F2}|$ = 5 mV Max.

Fig. 3-14 *CA3039 array device*

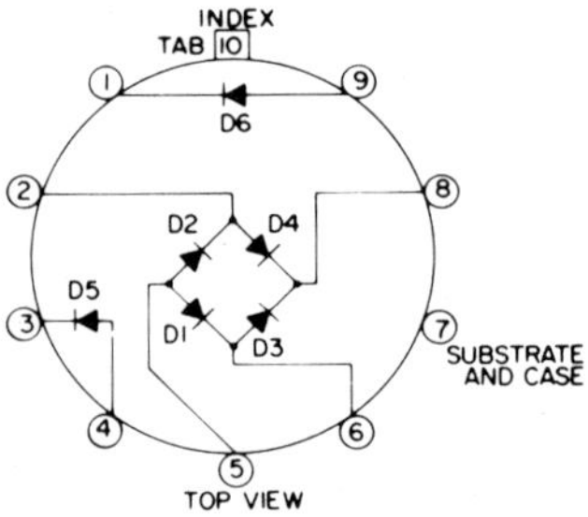

$V_{(BR)R}$ = 4V Min.
I_R = 10μA Max.
$|V_{F1} - V_{F2}|$ = 5mV Max.

Fig. 3-15 *CA3019 array device*

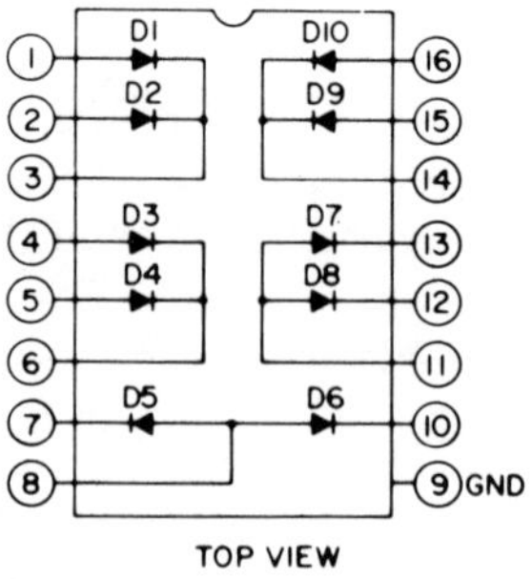

$V_{(BR)R}$ = 30V Min.
Current matching:
96% TYP.

Fig. 3-16 *CA3141 array device*

Part II

IC Applications Section

4

Voltage Regulator, Reference, and Power Supply Circuitry

In this chapter are discussed a wide variety of techniques that are useful in the stabilization of circuit voltage and current operating parameters. All circuits are ready to use as they are illustrated, but most are also readily adaptable to more diverse requirements as well.

Simple Voltage Regulators

In Fig. 4-1 are illustrated a group of voltage regulators that feature buffered, Zener-regulated outputs. The 3083 and 3093 array device types are illustrated because of their higher current capacity. The circuits can, of course, be adapted (by repinning) to the more general 3086 or 3046 units, if desired, but with a sacrifice in maximum current capacity.

The circuit of Fig. 4-1(a) is a Darlington-buffered 7-V regulator, which can handle load currents of up to 50 mA. The output voltage is that of the Zener, as the forward voltage of the Q1–Q3 emitter follower cancels that of diodes D1 and Q2. Load regulation is 2 percent or less, for a 40-mA current change, whereas line (input) voltage regulation is on the order of 0.2 percent per volt.

Options possible are higher input voltages—for which the 3183 will be found useful—or multiple outputs—by making Q1 and Q3 single-emitter followers. Note that the peak output current should be limited to protect against short circuits; thus, the (dotted) series collector

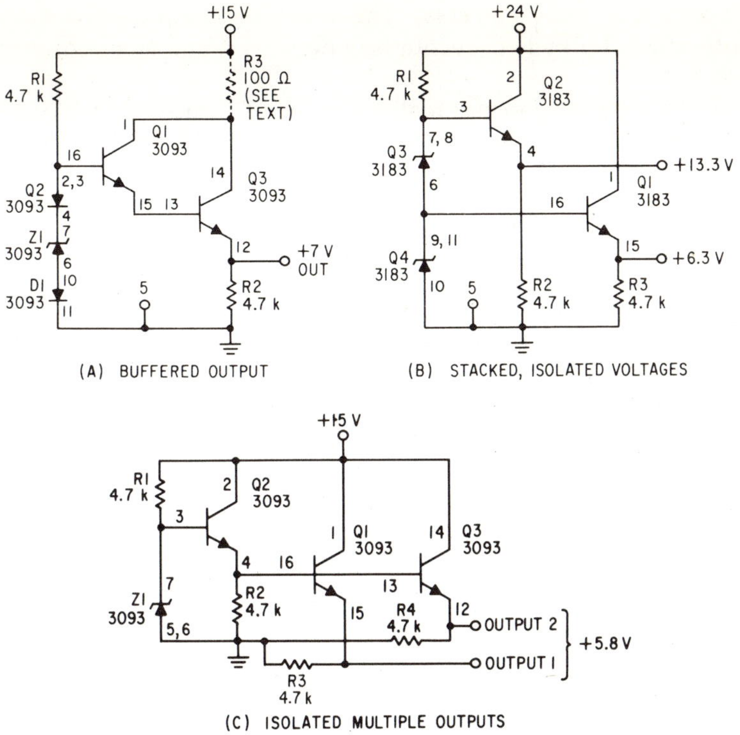

Fig. 4-1 *Simple voltage regulators: (a) Darlington-buffered output, (b) stacked and isolated voltages, and (c) isolated multiple outputs*

resistance (R3) is recommended. Such basic precautions extend generally, of course, to other IC array emitter-follower type circuits for obvious reasons of reliability.

In Fig. 4-1(b) is shown a variation in which stacked and isolated voltages are derived from a series Zener string, using the 3183 array. This variation allows buffered 6.3- and 13.3-V potentials to be generated, which can be useful, for example, in multistage amplifier circuits.

Possible variations of this circuit include more Zeners in the "stack" or Zeners combined with diode junctions to set output voltage to finer steps. Note that the 3146 is useful for this type of circuit as well if the load current requirements are only a few milliamperes.

Figure 4-1(c) shows a circuit that is useful where a very high degree of isolation or decoupling is necessary between circuit sections. Here, outputs 1 and 2 have the same nominal voltage level—5.8 V—but are fed from individual sources, Q1 and Q3. Thus, load currents that may change the voltage seen at output 1 will not substantially alter the voltage at output 2 (and vice versa). This relatively simple active circuit can far outperform bulky R-C decoupling networks, and at a saving in space and cost.

Highest performance is realized when Q2 is used to drive Q1-Q3 as shown, but the circuit is also effective with Q1-Q3 fed directly from the Zener (in which case the output is 0.6 V higher).

Voltage References

With the Zener diodes and active devices available in IC arrays, voltage reference sources capable of quite high performance can be built with relative ease. Voltage references are, of course, a type of voltage regulator or stabilizer, as we have been discussing. A voltage reference, however, has a much higher degree of overall accuracy than does a simple regulator, that is, the reference circuit has very low errors (a high degree of regulation) for line and load changes, as well as temperature variations. It is entirely feasible to build IC array circuits with errors caused by these sources that are well below 0.1 percent, as a matter of fact. This section discusses a variety of circuits useful for voltage references.

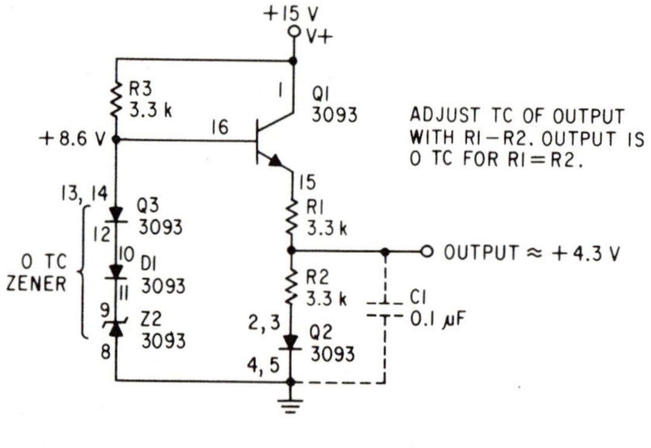

Fig. 4-2 *Adjustable TC reference (unbuffered)*

In Fig. 4-2 is shown an example of a low TC voltage reference circuit, patterned after Fig. 2-10. The circuit is called an adjustable TC reference because it does in fact allow the TC to be adjusted—to zero, if desired.

In operation, R3 sets up a current of 2 mA in the diode-Zener string, which functions as a (nominally) zero TC Zener diode, with a voltage of 8.6 V. Q1 buffers this zero TC voltage, but because of its V_{BE} TC, the voltage seen at the top of R1 has a positive TC. Diode-connected

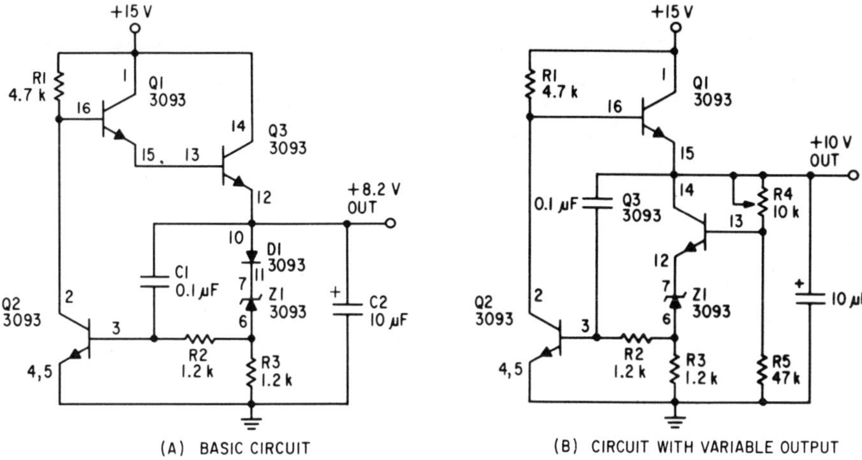

Fig. 4-3 Low TC reference with buffered output: (a) basic circuit, and (b) circuit with variable output

Q2 introduces a similar but negative TC. The voltages of Q1 and Q2 are summed by R1–R2, and their mid point (the output) will represent a zero TC voltage when R1–R2 and the positive and negative TC's are equal.

For the values as shown, the zero TC voltage resulting is 4.3 V, and minor adjustments in TC can be realized by trimming the relative values of R1 and R2. C1 functions as a noise bypass and is optional. Note that the output impedance of this circuit is high; actually, it is the parallel equivalent of resistances R1-R2. It must therefore be used in conjunction with a circuit that provides buffering or otherwise minimizes loading effects.

For a stable output, the current in the Zener should be kept constant, which in turn implies that the V+ voltage should be regulated. Alternately, if the circuit must be used with a varying source of voltage, a subsequent circuit can be used to stabilize the Zener current (and thus the output voltage).

Figure 4-3 illustrates a pair of circuits that provide low TC reference operation with a buffered output. The buffered output is achieved by means of a closed loop regulator, which maintains the voltage constant to within a few millivolts of the desired voltage.

In the circuit of Fig. 4-3(a), the output voltage of 8.2 V is established by the V_{BE} of Q2 and D1 and the Zener voltage of Z1. The Zener, as can be noted, is connected in a feedback loop around Q2 so that it tends to maintain the voltage at the emitter of Q3 constant despite the substantial current outputs from Q3.

R3 sets the current in the Zener at a level equal to the V_{BE} of Q2 divided by the value of R3. The series connection of two diodes and the Zener provides a low TC. C1 and C2 provide the functions of noise reduction and stabilization, respectively. Their values, not overly critical, may vary from those shown.

Performance of the circuit is quite good with regard to loading, for its output impedance is approximately 1 Ω because of the feedback. This means that a 10-mA load, for example, will change the output voltage only 10 mV, or about 0.1 percent. Line regulation is better than 0.1 percent/V.

The basic circuit of Fig. 4-3(a) is limited in the sense that there is no convenient means of adjusting the output voltage to voltages other than the basic Zener voltage. When output voltages higher than 8.2 V are desired, however, the variation shown in Fig. 4-3(b) will be useful.

In this circuit, the diode D1 of Fig. 4-3(a) is replaced by a transistor—Q3 in this case. Viewed from the base of Q3, this circuit is still an 8.2-V regulator, as previously. However, the voltage divider R4–R5 now allows the regulated output to be scaled up from 8.2 V by an amount set by the voltage divider. As shown here, a fixed R5 of 47 k sets up a current of 180 μA in the divider, and the pot R4 trims the output to 10 V. In most other regards, this version is like the basic circuit and has comparable performance (in terms of percentage of regulation).

High Supply Rejection Reference Circuits

Thus far, the circuits shown have only modest performance capability in terms of sensitivity to *input voltage changes.* The circuits shown in Fig. 4-4, however, have been optimized in this regard and operate over a wide range of inputs with very small output voltage changes.

In Fig. 4-4(a), Q3 of a 3096 array is connected as a Zener and fed a reference current from pnp Q4 of the same array. The current in Q4 is set up by Q2, the Zener voltage, and R3. Thus the Zener actually regulates its own current, which is relatively independent of the input voltage.

This description assumes that the circuit is already working, which in fact it will *not* automatically do at switch-on without outside aid. A start-up circuit consisting of Q1 and the divider R1–R2 forces a voltage across the base of Q2 that starts the current loop into operation. Q3 then breaks down at 7 V; because of the lower voltage across R2, this back-biases Q1. The back-bias locks out the start-up circuit and removes the coupling of input changes to Q3. This feature provides the immunity to input changes.

The circuit as shown will operate from 9 to 20 V and produces a 7-V output. Line regulation is quite good—about 0.02 percent/V. C1 serves as a noise bypass, reducing the noise generated by Zener action.

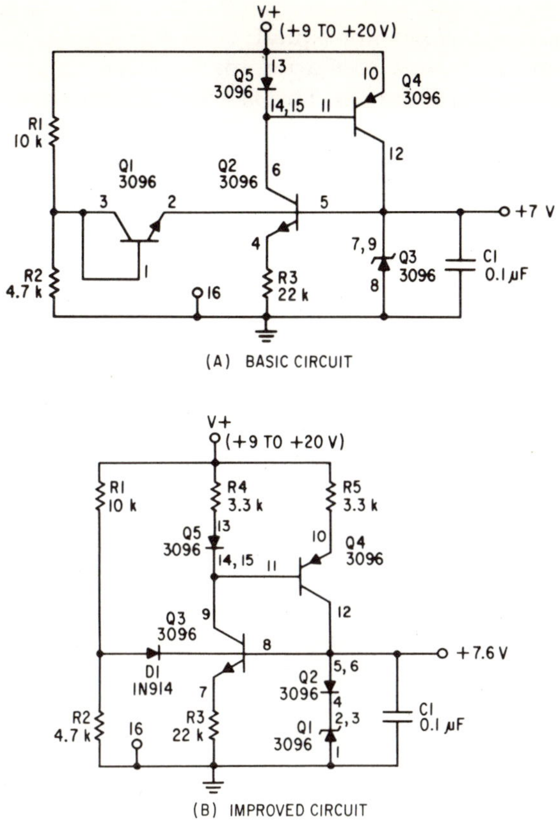

Fig. 4-4 *High supply rejection reference circuits: (a) basic circuit, and (b) improved circuit*

With some slight rearrangement of connections, the circuit can be improved even further in terms of line regulation and TC. This option is shown in Fig. 4-4(b).

In this circuit, the 3096 array is used once again, but a diode junction is added in series with the Zener for lower TC. To improve line rejection, emitter resistors R4 and R5 are added in series with current-mirror transistors Q4 and Q5. This simple change improves the line rejection to better than 0.01 percent/V. The circuit operates over the same range of voltages.

Since these two circuits are very useful in terms of producing a reference voltage highly independent of the input, some further options can be suggested. Using the 3096 array as well, *two* forward diodes could be connected in series with the Zener to provide a minimum TC connection. The function of Q3 (as shown here) could be performed by a single npn discrete (2N3904, or equal). Alternatively, one could simply use the npn devices from a 3086 array for the reference, the Q3 and D1 function, and a pnp discrete pair for Q4–Q5. Note that Q4–Q5 need not necessarily be matched when emitter resistors are used since the voltage drop across these resistors swamps out any V_{BE} difference. 2N3906 types (or equivalents) will work in this manner, as will many similar general-purpose types.

For both these circuits, loading of the reference voltage should be controlled for best accuracy.

A most interesting form of voltage reference from the standpoint of minimum TC is the type that stabilizes the IC substrate to a constant temperature above the highest expected ambient where transistor parameters can be held to a near minimum of variation with temperature. This means V_{BE} or V_Z variations with temperature can be reduced to negligible proportions, allowing very steady operation of voltage references and other such critical circuits. Of couse, other temperature-sensitive transistor parameters become stable as well, such as gain and bias currents. The technique is quite a useful one; its major drawback is (quite naturally) the extra circuit power required to heat the chip. Two circuits useful for the temperature stabilization of arrays are shown in Fig. 4-5.

In Fig. 4-5(a) is shown an interesting circuit by Dvorak that uses only a single transistor from a 3183 array to stabilize the array temperature.[1] The remaining four devices of the IC are available for application, as the designer wishes, as gain-stable transistors, diodes, or Zeners.

In operation, a fixed current is forced into Q1 via R1 from the −15-V supply. At an elevated temperature, the V_{BE} of Q1 will develop a voltage that is sensed by R2. Since V_{BE} is inversely proportional to temperature and the current in Q1 is fixed, the emitter voltage of Q1 serves as an accurate thermometer for substrate temperature.

The −15-V supply is reduced to the desired reference voltage by the R3–R4–R5 divider and applied to the reference input of a 741 op amp. This op amp operates as a high-gain differential amplifier and integrator. The output of the op amp drives the 2N1613 transistor, which adjusts the collector voltage of Q1; this action effectively adjusts the power dissipated in Q1. The closed loop thus formed applies power to Q1 until the V_{BE} equals V_{REF}, and thereafter smaller power increments so as to maintain this temperature. With the circuit values shown, the temperature of Q1 is maintained at 63°C.

Use of the 3183 (as shown) allows the remaining transistors of the array to be used over the full ±15-V power supply range. For such

[1] See N. Dvorak, Ref. (3).

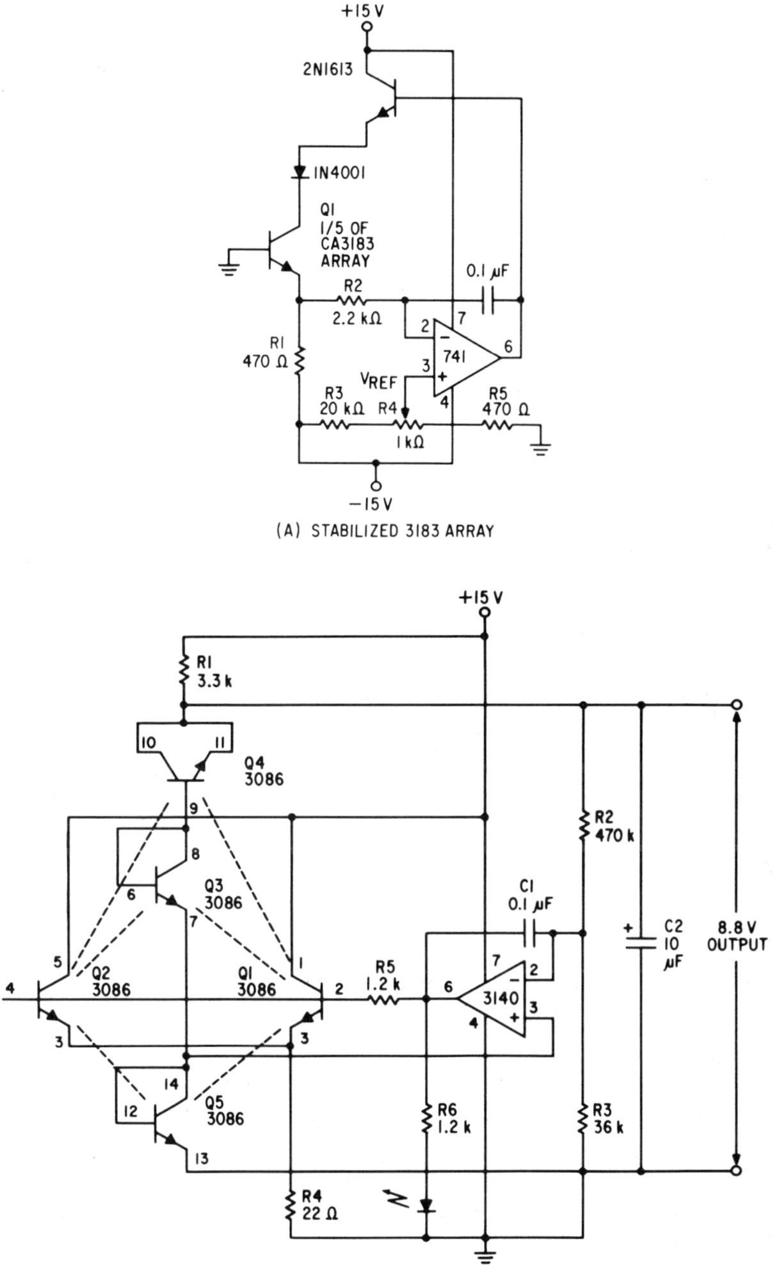

Fig. 4-5 *Temperature-stabilized circuits: (a) stabilized 3183 array, and (b) stabilized Zener reference*

applications, the substrate pin should be tied to −15-V. Note that a 3083 can also be used in this scheme (as could a 3093) if the V_{BE} seen by Q1 (and the other transistors) is maintained at 15 V or less.

Another circuit for stabilizing substrate temperature is shown in Fig. 4-5(b). This circuit operates from a single +15-V supply and uses a 3086 array. In this circuit, the scheme of sensing V_{BE} to monitor temperature is used, with Q5 performing this function. Q5, Q3, and Zener-connected Q4 form a low TC Zener, which has a reference voltage of 8.8 V at 2 mA of current. A portion of this stable voltage is tapped off by R2–R3, forming a reference voltage that is compared to the V_{BE} of Q5. The 3140 op amp compares these two voltages and drives Q1–Q2, which functions as the chip heater.

With the R2–R3 ratio shown, the chip temperature is raised to approximately 50°C—sufficiently high for use in moderate environments. About 25 mA of heater current is required in Q1–Q2 to maintain this temperature. This can be lowered somewhat if the 3086 array is covered with a foam-insulating shield and minimal heat sinking is used on its leads. Warmup time is less than 10 seconds, and the LED indicates thermal equilibrium (when out).

The 3086 devices, which are connected as shown and operated at a current of 2 mA, have a low TC, even without stabilization. For some samples tested, the TC over a 30°C range was less than 50 ppM/°C. When stabilized with the temperature regulator, this is reduced to 5 ppM/°C or less, or to a level where other errors such as long term drift and noise become dominant. The noise output for the circuit shown is reduced by the use of bypass capacitor C2.

Although this circuit is relatively immune to voltage output errors caused by temperature, it will still be sensitive to line-voltage changes unless the current in the reference Zener is kept constant. The high-line-rejection circuit of Fig. 4-4(c) can be adapted to solve this problem, if desired. Loading effects on the reference voltage output must be minimized as well.

The voltage reference circuit of Fig. 4-6 gets around the load and line sensitivity of the more common Zener reference circuits by using a reference based on the bandgap principle. When used with an op amp, the reference provides buffering against both line and load changes. This form of circuit is similar to the one described by Kuijk.[2]

In this circuit, the five transistors of a 3046 array are used to create a reference voltage of 1.205 V—the silicon bandgap voltage. Q1–Q4 and Q5 are operated at equal currents of about 100 μA. With Q1–Q4 in parallel, this causes the current density of this branch to be one-quarter that of Q5, developing a ΔV_{BE} across R3. The output is scaled to 1.205 V by

[2] See K. E. Kuijk, Ref. (2).

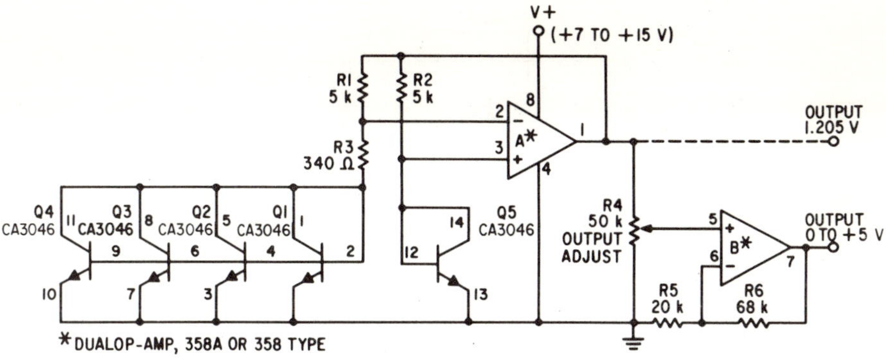

Fig. 4-6 Bandgap voltage reference

R1–R2–R3 and appears at the output of amplifier A. Note that a ground-sensing, single-supply op amp must be used here since 741 types will *not* function.

The buffered 1.205-V output can be used as shown, or section B of the op amp can be used to create a variable buffered source of 0 to +5 V, as shown. This circuit makes a handy dc voltage calibrator and can be battery operated if desired since its standby current is less than 1 mA. Since the line rejection of the circuit is typically less than 0.01 percent/V, varying supply voltages or aging batteries have little effect on the output.

Shunt Regulators

Although the series type of voltage regulator is the most common because of its efficiency advantages, shunt regulators are very useful for low-power regulators. They have the advantage of a high degree of decoupling that results from their low dynamic impedance. A Zener is a simple form of shunt regulator, of course, but the types we will discuss in this next section offer improvements over the simple Zener in terms of lower impedances, more adjustability, more stability, and so on.

A simple shunt regulator that illustrates these principles is shown in Fig. 4-7. This circuit uses a 3093 array and has a regulated output voltage of 8.6 V that is temperature compensated.

In this hookup, the Zener Z1 drives the base of transistor Q2, which in turn handles the bulk of the load current. The Zener current is set by the base resistor, which, as shown, sets up a 4-mA Zener current for low TC. This configuration is also sometimes called an "amplified Zener," for Q2 does in fact amplify the current capacity of the Zener.

Load regulation is typically 3 percent for a 40-mA current change, whereas line regulation is about 0.5 percent/V. The TC is less than 100

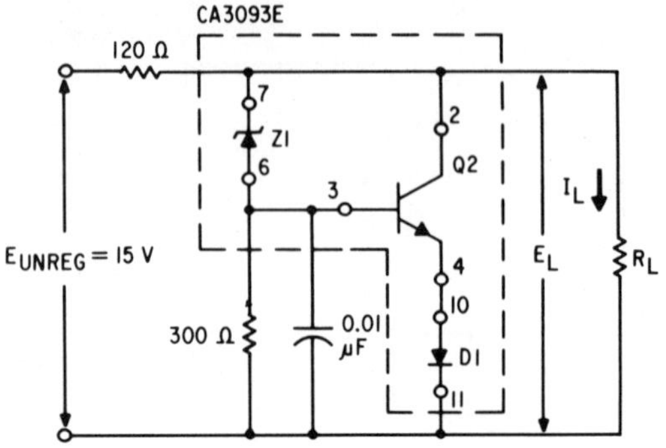

Fig. 4-7 Simple shunt regulator (Courtesy, *RCA Solid State Division)*

ppM/°C. Circuits of this type can be implemented with the 3083, 3183, and 3097 devices, as well as the 3093 illustrated. Possible modifications to this basic form are Darlington connections for Q2, paralleled units for greater current, or stacked regulators for higher voltages.

The shunt regulator of Fig. 4-8 illustrates how a Zener can be used to generate a *variable* shunt regulated voltage. This circuit uses the 3097, which contains a variety of active devices. Zener D1 furnishes a regulated 8 V across the output pot, whose arm feeds the pnp-npn composite. This circuit is the functional equivalent of a medium current pnp, providing regulation for currents of up to 40 mA with regulation on the order of 1 percent. The circuit is rated for operation over a 5 to 7.5-V range.

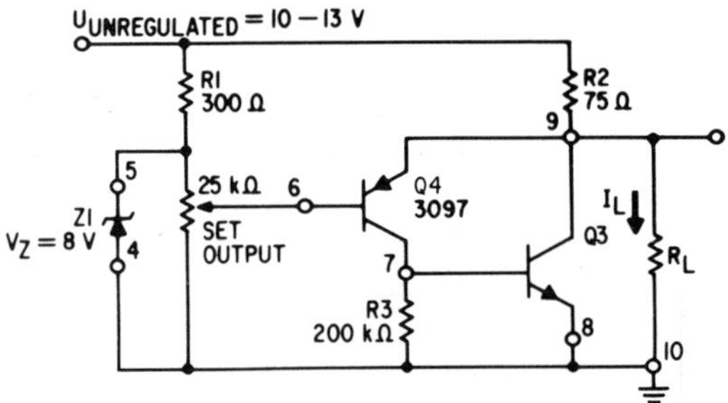

Fig. 4-8 Variable shunt regulator using the 3097 (Courtesy, *RCA Solid State Division)*

A "two-terminal" type of adjustable shunt regulator is illustrated in Fig. 4-9.[3] This circuit is somewhat similar to the adjustable series

[3] For source, see W. J. Jung, Ref. (10).

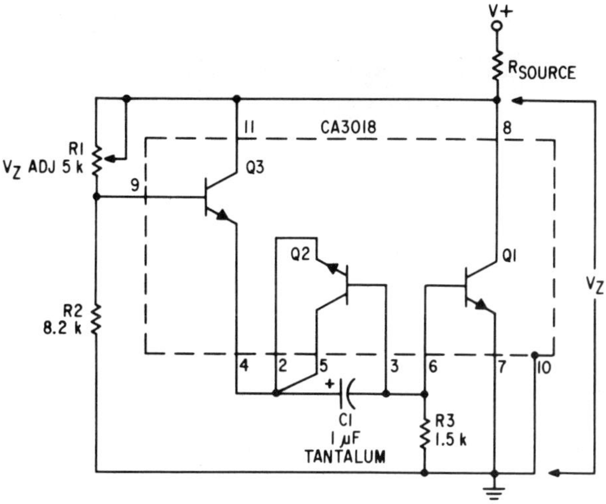

Fig. 4-9 Adjustable Zener shunt regulator

voltage regulator of Fig. 4-3(b). In this case, however, its function of operation is as a shunt regulator—in effect, an "adjustable zener."

The voltage at the base of Q3 under conditions of regulation is about 8 V, and R1–R2 scales this voltage upward. The scaling for the values as shown is for a 10-V output, but this may be changed if desired. The circuit can handle currents of up to 10 mA, or higher if repinned for a 3083 or 3093 array. Note that additional transistors can be paralleled with Q1, if desired, as an alternative means of current increase.

Figure 4-10 illustrates a useful variation on this circuit that increases its regulation capability quite substantially. In this circuit,

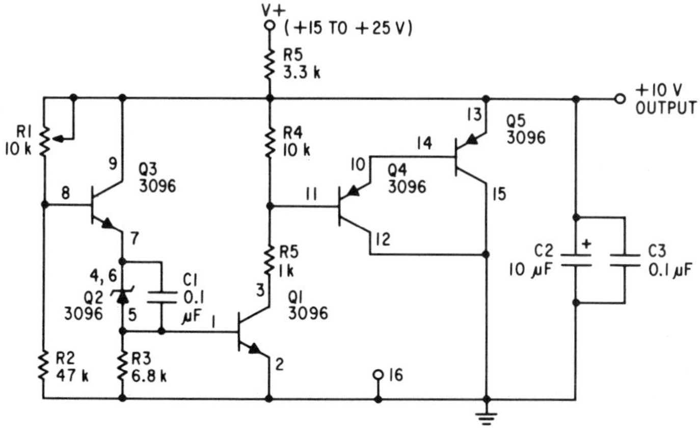

Fig. 4-10 Low-impedance shunt regulator

Q1–Q3 performs functions analogous to their counterparts in Fig. 4-9. In Fig. 4-10, however, Q1 does not drive the load directly but instead drives the Darlington pnp buffer Q4–Q5. Using the devices of the 3096 array, this circuit has a regulation of 0.05 percent for a 5-mA current change. Unfortunately, the phase shift and low gain of lateral pnp's Q4–Q5 prevent high currents. A discrete pnp device (such as the RCA8203) will not only get around this problem but allow currents of up to 1 A as well (if desired).

It should be understood, then, that the ultimate current rating of this circuit is primarily based on the transistors used for the Q4–Q5 function and that this rating is adjustable over a wide latitude from 10 mA to over 1 A. For Q1–Q3, any of the general-purpose(GP) array types such as the 3086 or 3018 may also be used (repinned, of course).

This circuit is also capable of reasonably low TC. For example, R3 can be adjusted to a value that minimizes the overall TC. If this goal is desired, values in the range of 300 to 1,000 Ω are suggested for R3, with R4 roughly twice the R3 value.

A very high-performance, high-current shunt regulator is shown in Fig. 4-11.[4] This circuit uses a very high-gain feedback loop for control and is capable of controlling ampere-level currents to within less than 1 mV, or less than 0.01 percent.

In this circuit, a 3018 IC array is used, with Q1–Q2 operating as a differential input pair and Q4–Q5 as a Darlington driver for the 2N3055 power transistor, Q6. Q3 is a small discrete npn used as a level shifter. D1 is a 6.6-V reference using a plastic transistor.[5] It can, of course, be an IC low-TC Zener connection, as well.

This circuit is actually a power op amp, set up as a shunt regulator at −12 V with a 1-A current capability. R10 adjusts the output voltage, which is scaled up from the basic reference voltage by the R9–R10–R11 divider. R1 sets the current in the Zener to the reference value for minimum TC. For the device shown, for example, the values would be 1 K, or a 5-mA current, in general.

To adapt this circuit to other voltage or current levels, several simple changes are necessary. A 3183, 3146, or 3118 array can operate up to 30 V, if desired. A 3183 (or 3083) is preferable from a *current* capability standpoint since it will allow more base drive for Q6. Note that it is possible to use the five npn devices of the 3183 or 3083 for the reference diode—Q1–Q2, Q4, and Q5—or the 3146 or 3046 in a similar fashion for lower currents. The 3183 or 3083 will allow several amps of current to be handled if R6 is taken to a lower value, such as 100 Ω.

---

[4] For source, see W. J. Jung, Ref. (7).
[5] See E. J. Kennedy, Ref. (11).

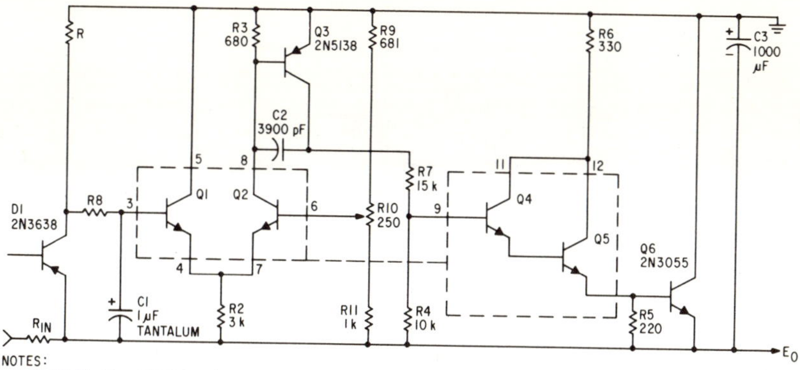

NOTES:
1. Q1, Q2, Q4, Q5 = CA3018 (RCA)
2. R8 = PARALLEL EQUIVALENT OF R9-R11

Fig. 4-11 *High-power shunt regulator*

Current Regulators

While the discussion thus far has stressed voltage regulator circuitry, there are also many occasions where regulation of current is desirable. Several circuits are illustrated in this section that are useful for this purpose.

Figure 4-12 illustrates a high-accuracy current source that produces an output current quite independent of supply voltage or load impedance as a result of the local feedback in the current mirror.

The circuit may be recognized as a variation of the circuit of Fig. 4-4(b), which also has high supply rejection. In this circuit, diode-connected npn transistor Q2 is used to drive pnp's Q4–Q5. The output of Q4 constitutes the regulated current, which follows the equation shown. The circuit as shown does not provide minimum TC, but this may be accomplished by using three additional diodes in series with Zener Q1. The circuit is useful at currents of 5 mA or less.

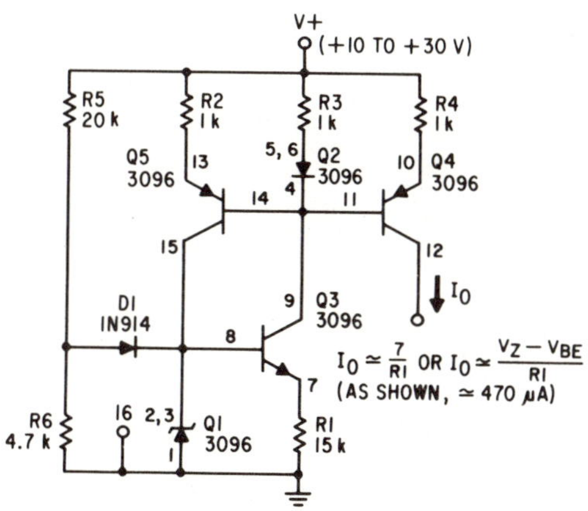

Fig. 4-12 *High-accuracy current source*

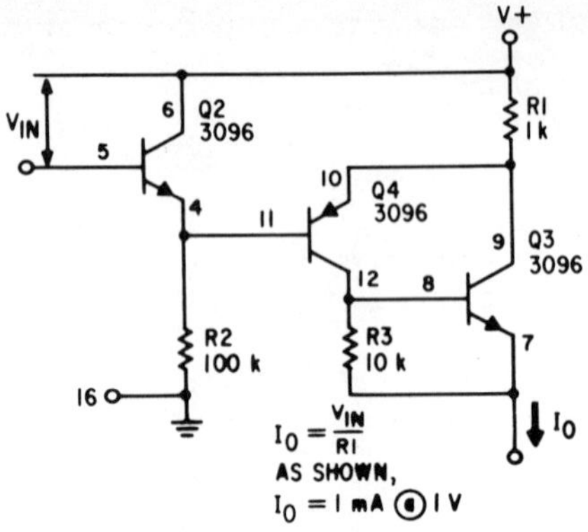

Fig. 4-13 *Voltage-controlled current source*

A voltage-controlled current source is shown in Fig. 4-13, which allows a current to be developed that directly follows a controlling voltage, V_{IN}.

This circuit uses the npn Q2 of a 3096 array to drive a pnp-npn composite, Q4–Q3. The respective V_{BE}'s of Q2 and Q4 essentially cancel one another, thus causing the input voltage V_{IN} to appear across R1. The output current is then simply equal to V_{IN} divided by the value of R1.

As the equation shows, R1 can be used to scale the voltage to current-control sensitivity—to match a given input voltage range to a desired (proportional) current range. This circuit has a wide range of uses in voltage-controlled timers, oscillators, and the like. Maximum current is about 10 mA, as shown.

The circuit of Fig. 4-14 is a precision, multiple-output, current-mirror circuit that uses a 3084 pnp array. Q5 and Q6, the Darlington-connected pair, drive D1 and Q4 to form a low-error current mirror. Q3, Q1, and Q2 are slaved to this mirror, which means that currents I1, I2, and I3 are equal to the input current, I_R.

A subtle trick to generate three matched currents can be played with emitter resistors R1 and R2 and R3 when it is desired to use emitter resistance to increase the output impedance of the current source and make it independent of the V+ voltage. Since R1 must conduct three times the reference current, it is made one-third the value of R2 and R3. This insures that all transistors will operate at equal current densities, thus minimizing drift and offset.

The circuit, which operates over a 10- to 40-V range of V+, is useful wherever multiple precision currents are required.

Lab Power Supplies

If we combine some of the array design techniques discussed thus far, we can with relative ease design high-performance lab power supplies. Two examples are illustrated in Fig. 4-15.

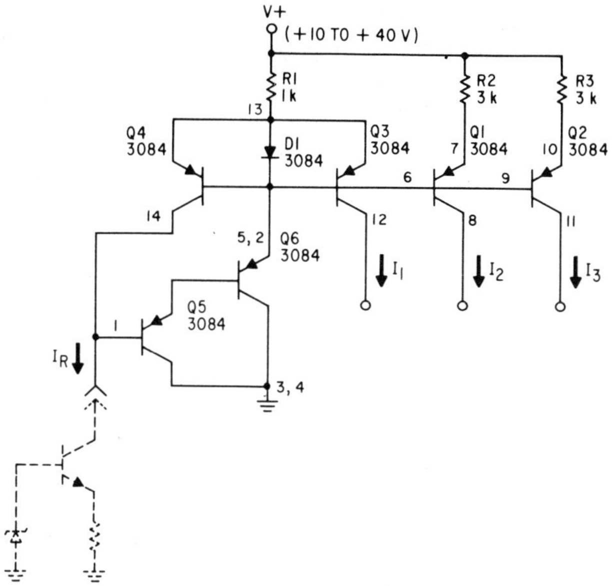

Fig. 4-14 *Precision, multiple-output, matched-current source*

A simple lab supply suitable for use over a range of 0 to 15 V at up to 50 mA of output current is shown in Fig. 4-15(a). This circuit is composed of three basic sections: a reference section made up of a 3086, a control amplifier using a 3140 op amp, and a "pass transistor" using a second 3086 array.

The first 3086 functions simply to produce a regulated 8.4 V, which is applied to R7, a pot used to adjust output voltage. A scheme of double regulation is used in this 3086, with Q3–Q4 preregulating to a 12-V level, and Q1–Q2 and Q5 providing a final regulation—to 8.4. This provides a line-regulation figure of 0.01 percent/V and a very low noise, 1 mV or less. Furthermore, this array operates at a fairly low TC.

The output voltage of the reference circuit is picked off by R7, which applies a variable voltage to the (+) input of the op amp. This op amp is a single-supply type; therefore, the voltage of 0 to 8.4 V from R8 is linearly scaled up by R9–R10 to 0 to 15 V at the (+) supply output terminal. R8 and C3 form a noise filter that reduces reference noise to a minimum.

The output series control function is provided by the second 3086. Here, four transistors are paralleled for greater current capability, allowing a 50-mA output. Current limit control is provided by Q5 of this array, which senses the current in R12–R13. When the voltage across this resistance reaches the V_{BE} of Q5, Q5 turns on and reduces the drive to Q1–Q4 to a level sufficiently low so as not to exceed the desired current-set by R13.

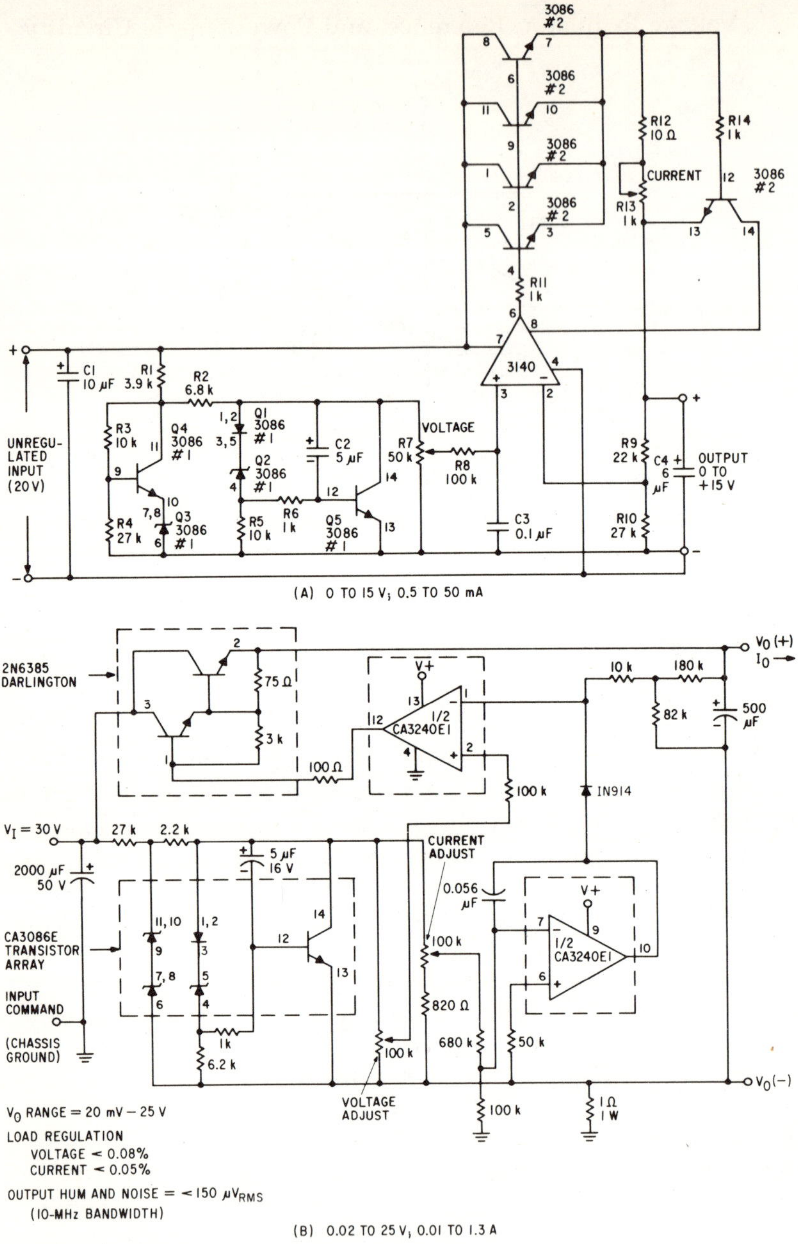

Fig. 4-15 *Lab power supplies: (a) 0 to 15 V, 0.5 to 50 mA; and (b) 0.02 to 25 V, 0.01 to 1.3 A* (Courtesy, *RCA Solid State Division*)

This circuit is capable of quite high performance within its operating limits and is functionally very useful because of its ability to adjust to zero output. Higher output current can be attained by using a 3083 array instead of the second 3086, if desired, or an npn Darlington (see next section).

The circuit of Fig. 4-15(b) is quite similar to the basic version of Fig. 4-15(a) in many regards. It uses a slightly different arrangement to achieve a performance of a 0- to 25-V range, with a current capability of up to 1.3 A. Both voltage and current are adjustable to precise levels.

In this circuit, the 3086 array supplies a reference voltage to the 100-K voltage adjust control. This is somewhat similar to the scheme used previously except that a different preregulator hookup is used. The output voltage is scaled up to a 0- to 25-V range by the output sense resistors (180 and 82 k), which provide negative voltage feedback to the upper half of the 3240 dual op (a dual version of the 3140).

The lower half of this op amp is connected so as to monitor the current in the 1-Ω, 1-W sample resistor. When the voltage dropped across this resistor is equal to that at the 680- to 100-k resistor junction, the current control loop will come into play. This occurs by the output at pin 10 of the op amp rising, thus turning on the IN914 diode and over-riding the voltage control loop. The net effect of the current control is to limit the output current to a maximum of the set by the 100-k current control. If load current demands are less than this current level (set over the 1.3-A max range), the voltage control loop will function in a conventional fashion.

The output current capability of this device is 1.3 A. A power Darlington pass transistor is used, and a heat sink is required.

REFERENCES FOR CHAPTER 4

(1) Widlar, R. J., "A Versatile Monolithic Voltage Regulator," National Semiconductor, AN-1, 1967.

(2) Kuijk, K. E., "A Precision Voltage Reference Source," *IEEE JSSC,* June 1973.

(3) Dvorak, N., "One Transistor Senses, Heats in Temperature Regulator," *Electronics,* March 18, 1976.

(4) DeAngelis, D. P., and Palumbo, M., "Regulator Holds Temperature of Chip's Substrate Constant," *Electronics,* Feb. 2 1970.

(5) Shah, M. J., "Using Transistor Arrays for Temperature Compensation," *Electronics,* April 12, 1973.

(6) Jung, W. G., "IC Array As An Adjustable Zener," *Electronic Engineer,* Sept. 1971.

(7) Jung, W. G., "Don't Shun the Shunt Regulator," *Electronic Design*, July 15, 1969.

(8) Jung, W. G., *IC Converter Cookbook*, Howard W. Sams and Company, 1978.

(9) Jung, W. G., *IC Op Amp Cookbook*, Howard W. Sams and Company, 1974.

(10) ______, "101 Analog Designs," Interdesign, Inc., Sunnyvale, Calif., 1976.

(11) Kennedy, E. J., "Inexpensive 6-V Reference Is Also Temperature Stable," *Electronic Design*, Nov. 8, 1967.

(12) Manufacturer's data sheets (RCA), 3018 (Transistor array), 3046 (Transistor array), 3086 (Transistor array), 3083 (Transistor array), 3093 (Transistor/Zener diode array), 3096 (Transistor array), 3097 (Thyristor/transistor array), 3183 (High-voltage transistor array), 3140 (Op amp), and 3240 (Dual op amp).

5

Amplifier Circuit Techniques

This chapter focuses on circuits that use IC array devices and are useful for signal amplification. This topic is indeed so broad that a detailed discussion might fill an entire book. For this reason, the intent is to cover the basic techniques rather than the variations possible. In conjunction with the design information of Chap. 2, the circuit ideas of this chapter can be one of the most rewarding subjects of this book.

Wherever possible, the circuits presented are useful working entities in themselves and also demonstrate the principles under discussion. Since the most fundamental principles of transistorized amplifier circuitry are probably already familiar to most readers, the discussion will be directed toward explaining how IC arrays can enhance the value of a particular circuit.

Emitter Followers (Common Collector Amplifiers)

The emitter follower (or common collector stage, as in textbook usage) is a circuit workhouse so standard that it seems like overkill to discuss it at all. Several recurring problem areas encountered by emitter followers, however, can be dealt with effectively by using arrays. We will discuss these in this section.

A classic design problem associated with the emitter follower is how to realize enough current gain to buffer adequately the load being driven, that is, to attain an adequately high input impedance.

With multiple transistors readily available in arrays, this problem is greatly lessened since a Darlington emitter follower can be used to provide megohm input impedance with loads down to as low as 100 Ω.

Figure 5-1 illustrates two Darlington stages, using 3018 transistors (others are usable if desired.) Figure 5-1(a) is a straightforward example of a Darlington buffer, with Q3–Q4 operated at a 5-mA current for

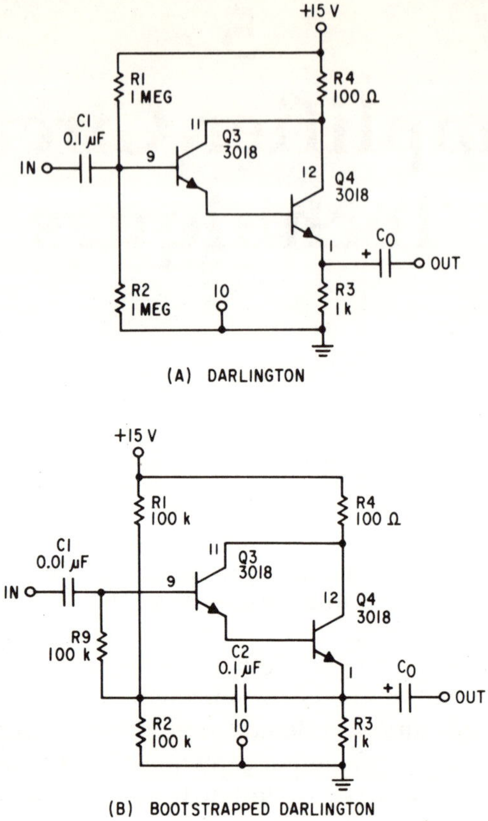

Fig. 5-1 *High-input-impedance emitter followers: (a) straightforward Darlington, and (b) bootstrapped Darlington*

moderate-load drive capability and to maintain high gain. At the 5-mA current level, the Darlington pair will have a typical gain of 7,000; thus, a 500-Ω load will appear as greater than 3 MΩ at the input. The input impedance of this circuit is conservatively set by R1–R2, and the bias resistors are set at 500k. R4 provides output short-circuit protection, and C_0 is an output coupling capacitor, chosen in value to satisfy a low-frequency rolloff requirement in conjunction with the input impedance of the *following* stage.[1] For example, 1 μF into 10k will result in a lower 3-dB point of 16 HZ.

With only a slight modification, the bootstrapped Darlington of Fig. 5-1(b) achieves an input impedance of several megohms. As was

[1] Similar capacitors are found in other circuits of the chapter.

noted in Fig. 5-1(a), the necessary bias network resistances can reduce the input impedance of a Darlington stage because they shunt the input.

In Fig. 5-1(b), capacitor C2 is added to drive the R1–R2 junction at the same level as the output. This reduces the signal voltage across R5 and raises its effective value to 10 MΩ or more. The input impedance of this circuit will then be limited by the Darlington current gain and the load—to a value of several megohms, typically.

The supply voltage for both circuits is indicated as +15 V, but it can deviate substantially without performance deterioration. For higher voltages, a 3118 or 3146 type device should be used, of course. For load currents above 5 to 10 mA, the 3083 (or 3183) is appropriate.

Low Power Biasing

With both the Darlington and standard single-stage emitter follower, the simple resistor pulldown (R3 of Fig. 5-1) has a couple of drawbacks. First, it tends to be wasteful of power, as this resistance must be driven as well as the load. It also restricts the linear dynamic range, because for high-level output swings, the current in Q4 will vary radically. Finally, for a given value of maximum load current, it can be shown that a constant current emitter bias consumes *less* standby power than a pulldown resistor.

The solution to these problems is the modified emitter follower of Fig. 5-2. Here, Q2 is a constant-current dc emitter feed for Q3–Q4, in this case set up for a current of 1 mA. Since Q2 is diode-biased by Q1,

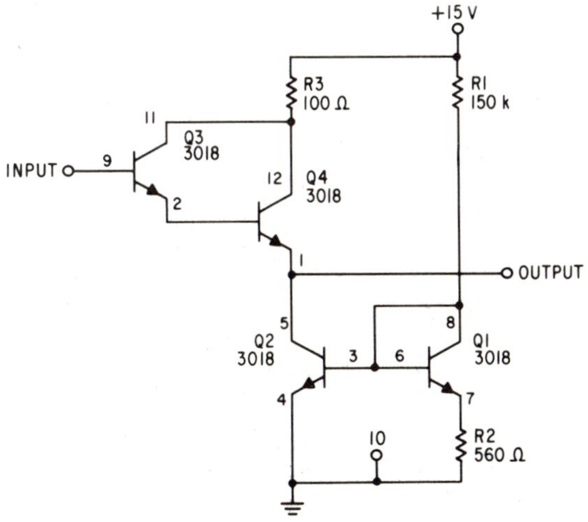

Fig. 5-2 Darlington emitter follower with low-power biasing

the 1-mA current is maintained down to very low output-voltage swings of Q4—to 1 V or less.

A further reduction in power consumption is achieved in Q1 by operating Q1–Q2 as a logarithmic current source. A reference current of 100 μA in R1 and an R2 value of 560 Ω biases Q2 to a 1-mA level. Were a straight 1/1 current source used, 900 μA of current would be wasted. This bias-current reduction technique can be applied to a great many such current sources; it is not limited to the particular application of an emitter follower.

Adjustable and Zero Offset Emitter Followers

It is often very useful in signal-processing circuits to have a unity gain buffer in the form of an emitter follower with zero input-output dc offset. Two schemes that accomplish this are shown in Fig. 5-3.

In Fig. 5-3(a) is shown an npn emitter follower that is capable of wideband operation because of the exclusion of slow pnp devices from the signal path. Q2 is the output buffer transistor, which is driven by matched device Q1, diode-connected. Q5 is a current source of a nominal current equal to that in Q2. With equal currents in Q1 and Q2, their V_{BE}'s are equal, and there will be zero dc offset.

As can be noted from the figure, resistor string R1–R2 biases both Q1 and Q2, through both the Q4–Q5 current mirror and Q3. The current-mirror "gain" is made variable by R4, which adjusts the relative currents of Q1 and Q2, providing about a ±40-mV variable dc offset.

An alternative scheme for a zero (or adjustable) offset follower is shown in Fig. 5-3(b). Here, pnp Q4 drives Q3 in a more conventional fashion but with some sacrifice in speed (because of Q4). Q3 and Q4 are biased to a current near 1 mA, and their V_{BE}'s are nearly equal.

The pot R2 trims the input-output offset to zero; alternatively, Q4 can be directly coupled to Q3 and the circuit will have a 30-mV offset because of the difference in npn and pnp V_{BE}'s (see Chap. 1).

Fixed Offset Emitter Follower

Just as zero offset emitter followers are desirable, so also is a fixed and predictable dc offset. Figure 5-4 shows how an offset of 8 V can be introduced in a signal path, with buffering at both input and output. Such a circuit is useful for level shifting, for example.

In this circuit, Q1 is an input buffer that drives Q2 connected as a Zener. Q2 in turn drives Q3, an output emitter follower. Base emitter resistor R4, in conjunction with the V_{BE} of Q3, forces the Zener current to be 2 mA, for minimum TC ($V_Z + 2V_{BE} = 0$ TC at 2 mA.) This results in a stable input-output offset. Q5 forms a current source feed for Q3, biased to a 5-mA level.

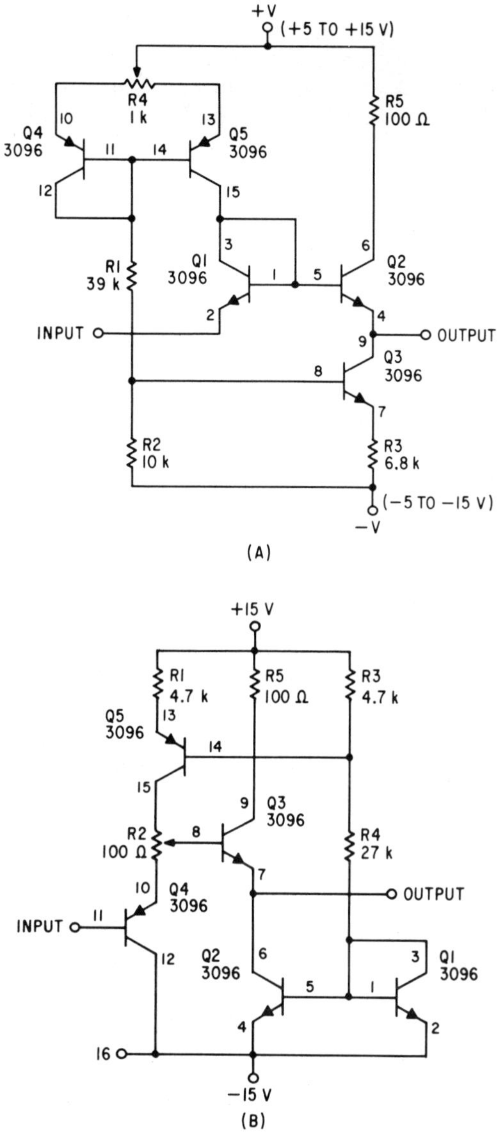

Fig. 5-3 *Adjustable-offset emitter followers: (a) npn emitter follower capable of wideband operation, and (b) alternative scheme*

This circuit can be used in a dc-coupled circuit where a fixed, buffered offset is desirable. The offset voltage can be incremented upward, if desired, by adding one (or more) diodes in series with the Zener.

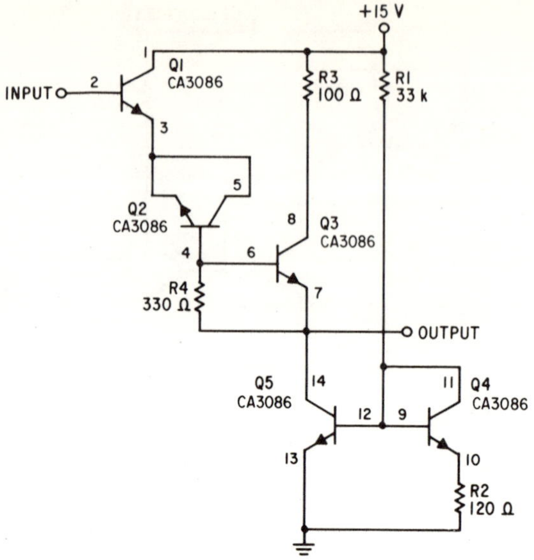

Fig. 5-4 *Fixed-offset emitter follower*

Line Driver

A circuit that is useful when a specific and accurate source impedance must be presented to the load is shown in Fig. 5-5. This is a line-driver circuit with local feedback. It is designed to drive video or audio lines.

In Fig. 5-5 is shown a version of the line driver suitable for video frequencies up to 10 MHz. A 3018 IC is used, and the circuit will deliver 1 V pp into a 75-Ω load. The circuit provides a matched 75-Ω source to a 75-Ω transmission line by virtue of the source termination R_o.

The circuit is a Darlington connection of Q3–Q4, with feedback applied around Q4—via Zener-connected Q1—and Q2, which functions as a dynamic current sink. The stage operates class A with a standing current of about 20 mA, set by R5. Because of the wide band of the circuit, input capacitance should be minimized, the layout compact, and the supplies well bypassed for RF.

Common-Base Amplifier

In Fig. 5-6 is shown a form of common base amplifier that is uniquely suited to IC transistors. The circuit is called a "diode-biased" common base amplifier because a monolithically matched diode-connected transistor is used to establish the bias of the common base transistor.

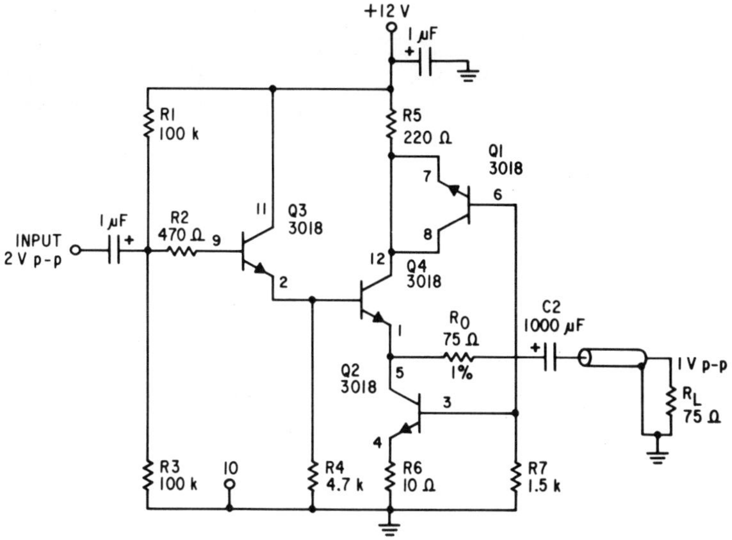

Fig. 5-5 Line driver

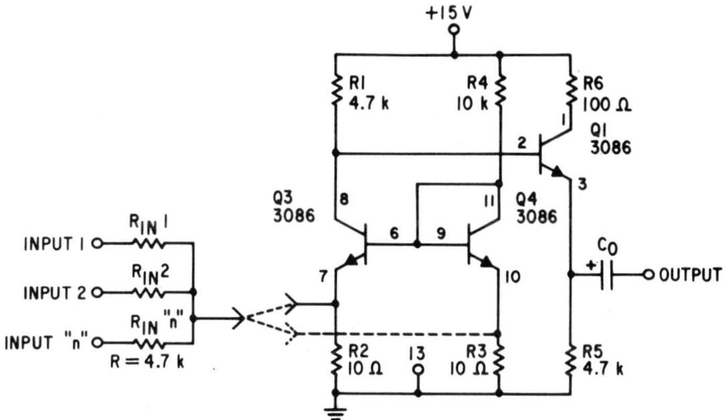

Fig. 5-6 Diode-biased base amplifier

In Fig. 5-6, Q3 is the common base amplifier and Q4 the bias diode. R4 sets up a dc bias current in Q4. With emitter resistors R2 and R3 equal, Q3 will conduct the same current as Q4, here 1.5 mA. Since R2 and R3 are small in value, they develop relatively small voltage drops. With R1 selected as one-half the value of R4, it will drop one-half the voltage and bias the collector of Q3 for optimum ac signal swing.

A characteristic of the common base amplifier is a low input impedance; in this case, the impedance at the emitter of Q3 is less than 10 Ω. Signals from low-impedance sources (voltage sources) are applied through series input resistors, such as R_{IN}, and the voltage gain of the circuit will be the ratio of R1 to R_{IN}. When R_{IN} equals 4.7k, the gain is unity; when R_{IN} equals 470 Ω, it is 10; and so forth. Note that a *number* of signals can be mixed at the emitter of Q3 without interaction because of the low input impedance. The polarity relationship is noninverting from Q3 to the output, and Q1 provides a buffered low output impedance source.

This circuit is capable of very wide bandwith if low resistance values and substantial bias currents are used because of the low parasitic capacitances and wide bandwidth of the monolithic process. Also, operation of Q3 in a common-base mode removes the bandwidth-reducing effect of collector base capacitance.

This circuit can also be operated as a *common-emitter* inverting amplifier if desired simply by applying the signal (or signals) to the emitter of *Q4*. This couples the signal through Q4 to the base of Q3, which is then driven in a common emitter fashion. The voltage gain operated in this fashion will be essentially the ratio of R1 to R_{IN}, but the signal polarity is inverted.

Feedback Summing Amplifier

By combining the natural current-summing properities of the common base connection with feedback, it is possible to build summing amplifiers that use IC arrays and offer very high performance. Such a circuit is the feedback summing amplifier of Fig. 5-7.[2]

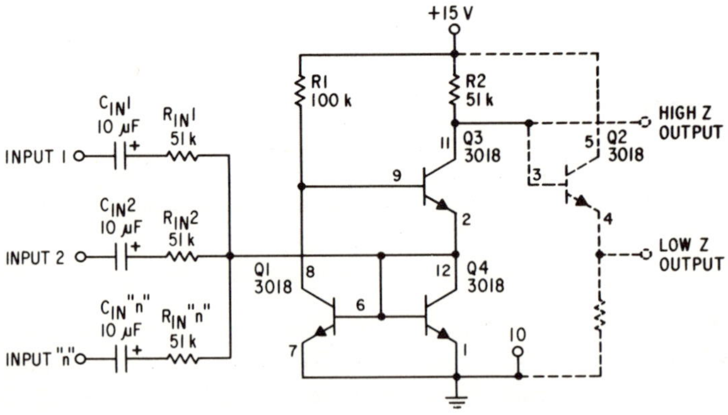

Fig. 5-7 Feedback summing amplifier

In this circuit, three transistors from a 3018 array are connected as a Wilson current source (see Chap. 2), biased at about 140 μA in Q1 and

[2] For source, see W. G. Jung, Ref. (10).

Q3–Q4. Because of the current feedback loop around Q3 and Q1–Q4, the impedance seen at the emitter of Q3 is very low, on the order of 3 Ω.

With 50-k (nominal) input or summing resistors, the circuit can be used as a low-distortion (≤ 0.1 percent), high-isolation (≥ 120 dB) audio mixer. Gain from any input (the number of which is unlimited) to the output across R_L is unity, as shown, but can be raised by lowering R_{IN}.

If a low output impedance is desired, the fourth array transistor (Q2) can be used as an emitter-follower buffer (shown dotted). For further information, see the reference cited.

Wideband Common Emitter Amplifier

A simple form of common emitter amplifier that is capable of wide bandwidth performance is shown in Fig. 5-8. This circuit uses the local feedback of the current mirror diode configuration as its basis of operation.

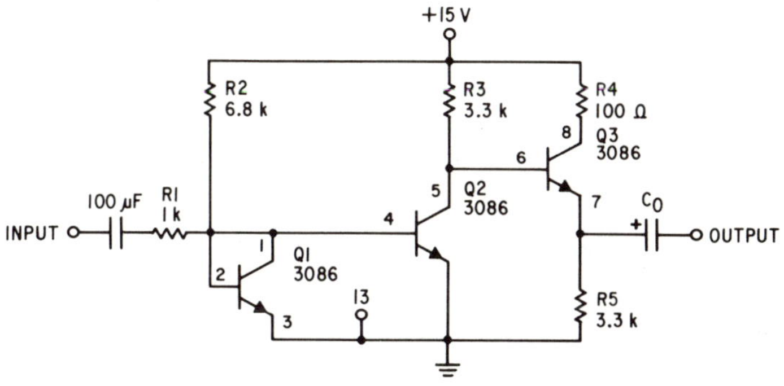

Fig. 5-8 Wideband common emitter amplifier

Q1 and Q2 of a 3086 are connected as a conventional current mirror, biased to a current of 2 mA by R2. R3, half the value of R2, biases the collector of Q2 for optimum ac signal swing.

If the input resistance, R1, through which an ac signal is applied is large compared to the diode impedance of Q1, the input signal current developed in Q1 will be determined essentially by R1. Just as the dc currents of Q1 and Q2 are duplicated, so also are the ac currents in Q2 and Q1 (and thus R3 and R1). Since the signal current in R3 and R1 is the same, the voltage gain of the circuit is the ratio of R3 and R1. The signal polarity is inverted because of the common emitter connection of Q2. Note that when R1 > R3, the circuit attenuates (if desired).

The circuit is useful over a wide range of frequencies as shown, and the value of R1 yields a gain of 10 dB. Gain can be increased by lowering R1, but with an increase in distortion. The distortion for the values shown is less than 0.3 percent for a 3-Vrms output level. The value of R1 is the input impedance of the circuit.

Simple Common Emitter Amplifier

A more conventional form of common emitter amplifier is the simple amplifier of Fig. 5-9. This circuit is very much like that of a common emitter amplifier designed with discrete transistors, with the exception of the bias technique used.

In this circuit, Q2 is the amplifier transistor, and Q4 is used to establish a matching bias. Since Q4 and Q2 are on the same chip, their dc currents will be equal when their base emitter voltages are equal. Here, equal value emitter and base resistances provide this relationship, with both base resistors fed from a common point, the collector of Q4. The desired emitter current is forced in Q4 by R1, which then causes Q2 to conduct an equal current. The value of R6, nominally one-half that of R1, causes the collector of Q2 to bias quiescently at one-half the supply voltage.

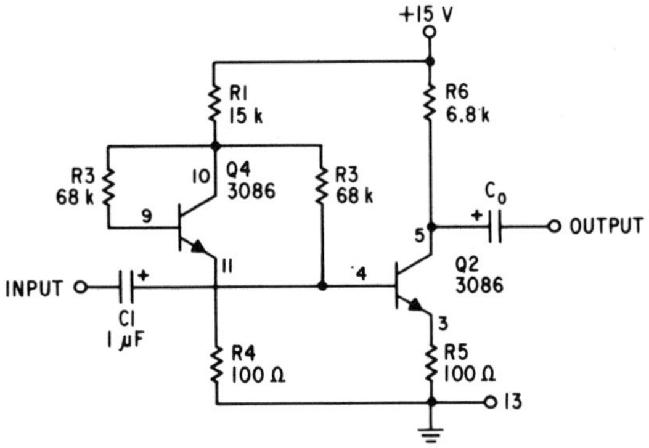

Fig. 5-9 Simple common emitter amplifier

Because of the dc bias conditions described, the ac gain of the circuit cannot be varied completely independently of bias considerations. As shown, the gain is about 36 dB; in general, it is set by the ratio of R6 to R5. Output loading must be minimized to realize this gain, of course. For low values of loading, an emitter follower buffer will be useful.

To modify the stage for other values of gain, the simplest approach is to fix R6 and vary R5 as desired. For example, gain becomes 20 dB (or 10/1) when R5 is 680 Ω. This approach will not disturb the dc bias conditions if the value of R4 is changed as well to maintain the same current in Q4 and Q2.

Input impedance of the circuit is fairly low, approximately 100 times the value of R5 (or 10k as shown). Input capacitor C1 must be selected for adequate signal coupling into this impedance, in shunt with R3.

This circuit is best suited for audio frequency applications, because its bandwidth is relatively restricted as shown as a result of the effect of collector-base feedback capacitance around Q2.

Cascode Amplifiers

A simple means of removing the bandwidth limiting restriction of the collector-base capacitance associated with a common emitter stage is to operate it as a *cascode* amplifier instead. The cascode amplifier, or, more specifically, a common-emitter/common-base cascode, attains the high-voltage gain of the common emitter stage, but with the wide bandwidth of the common base stage. Two examples are shown in Fig. 5-10.

In Fig. 5-10(a) is shown an example of a cascode amplifier designed for both high gain and bandwidth. Q2 and Q3 form the cascode amplifier, with dc biasing for this pair provided by Q1. A dc-bias current is forced in Q1 by R4, and Q2 is forced to conduct a similar current, since Q1 and Q2 are fed through equal value resistances from the collector of Q1. Since Q3 is in series with Q2, it conducts the same current as Q2. R3 is the collector load for the Q2–Q3 cascode.

Signalwise, the Q2–Q3 pair has the isolation of a common-base stage, but the input impedance and gain of a common-emitter stage. As shown, the gain is about 46 dB (200/1) and the bandwidth, approximately 3 to 4 MHz. For maximum bandwidth, an output buffer is desirable, and the layout should minimize capacitance at the Q3 collector.

Because the stage is effectively running "wide open" with no feedback, linearity of the stage is poor, particularly at high signal levels. For best linearity, input levels should be confined to levels of 10 mV pp, or less.

Alternatively, feedback can be introduced into the emitter of Q2 to improve linearity in the form of emitter resistance.

To provide a fixed or selectable gain cascode amplifier, or a high degree of linearity, the circuit of Fig. 5-10(b) can be useful. This circuit has the wide bandwidth properties of the cascode connection, but with an easily controlled gain.

The circuit can be recognized as an adaptation of the circuit in Fig. 5-8; in this case, Q1 drives Q5 in a similar fashion. The main difference is that Q3 is cascode-connected to Q5, and the signal is developed across R3 in the collector circuit of Q3. A tap on the R2–R6 bias string serves to fix the base of Q3 in the common-base mode.

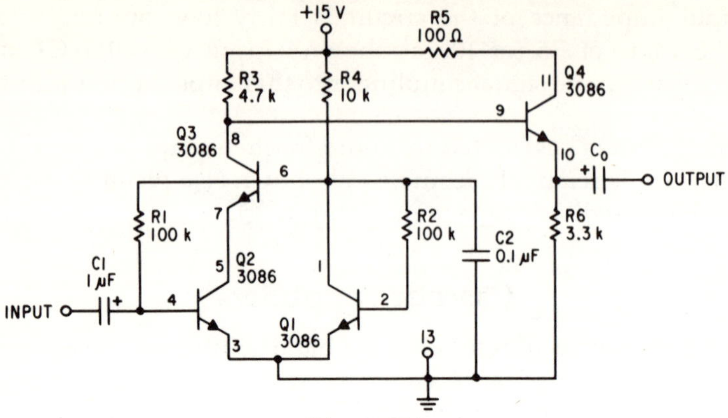

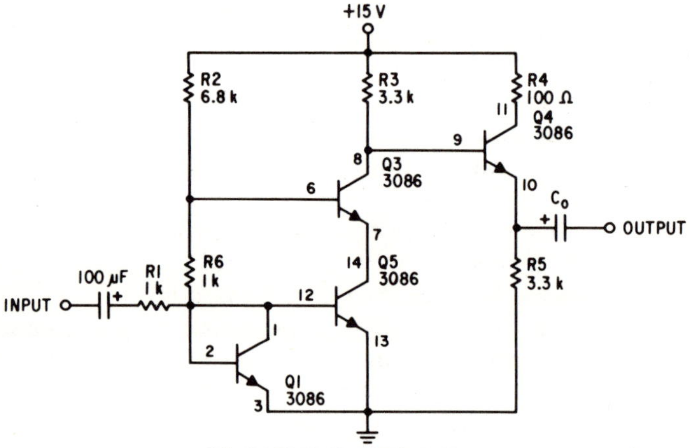

Fig. 5-10 *Cascode amplifiers: (a) high gain, and (b) fixed or selectable gain*

Gain of the circuit is set by the ratio of R3 to R1, as previously, and the circuit has distortion properties similar to those of Fig. 5-8. Bandwidth is best maintained by an output buffer, as shown, and low stray capacitance.

Feedback Amplifiers

There are a variety of feedback techniques that can be used with single-ended ac amplifiers using npn IC array transistors. The flexibility and performance capability of the devices allow stable and predictable amplifiers to be designed that are usable over a wide range of gains, bandwidths, and operating levels. Unlike the circuits just discussed, these circuits use moderate to heavy degrees of overall feedback to make performance largely dependent upon resistor ratios. Thus, some may be used as "gain blocks," with performance programmable by resistors or other external components.

The general-purpose amplifier of Fig. 5-11 is a good example of how feedback can be used to advantage to stabilize operating levels. In this circuit, Q1–Q3 compromise a feedback pair, with Q5 serving as a bias regulator and Q4 as an output buffer [see Reference (7)].

The design of this circuit can be much simplified if a few assumptions are made at the output. One of these is that the collector current of Q3 is ten (or more) times that of Q1. This allows some latitude for the value of R2, which in part determines the gain.

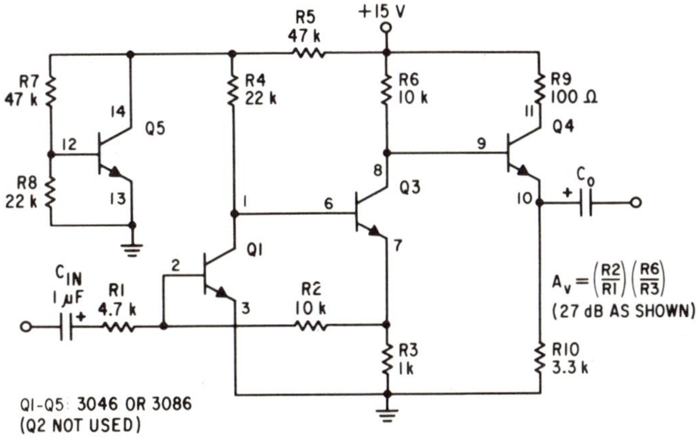

Fig. 5-11 General-purpose amplifier

Bias regulator Q5 produces a potential of 3 V_{BE} at the top of R4. From the dc feedback loop around Q1 and Q3, it can be noted that R3 will drop 1 V_{BE}, and 2 V_{BE} will appear at the base of Q3. Therefore, R4 will have 1 V_{BE} across it. The value of R4 then simply sets the collector current of Q1 as V_{BE}/R4. Since the V_{BE} of Q1 appears across R3, the Q3 collector current (flowing in R3) is V_{BE}/R3. R6 is chosen to drop about one-half the supply voltage when the collector current of Q3 is flowing. In this circuit, the current of Q1 is 30 μA, and in Q3, it is 600 μA.

The gain of the circuit is the product of two-ratios, R2/R1 and R6/R3. Of these values, only R1 has no effect on bias. R2, however, has only a small effect on bias and can range up to about 30k. As shown, the closed loop gain is 27 dB (or 21 times), and the input impedance is 5 k, or the value of R1 in general. Maximum output from the circuit is approximately 3 V rms, and loads down to about 3k can be driven.

Low-Noise Amplifier

For very low-level signal-amplification tasks, a preamplifier specifically designed for low-noise characteristics is desirable. With attention to appropriate design details, certain IC arrays are capable of excellent noise performance, particularly so when operated from low-resistance sources. A design suitable for use in applications where low noise is important is shown in Fig. 5-12.

To minimize amplifier noise, a number of rules of thumb, correctly applied, will bear fruit. One of these is to minimize the resistances seen in the signal (or feedback) path, because resistances in themselves generate noise. Another is to operate the first stage at a relatively high current, which minimizes its input *voltage* noise. Also, the gain of the first stage should be relatively high to make the noise of the second stage negligible. If the amplifier is a feedback type, a noninverting connection (also called series feedback) is preferred over an inverting one, as it is inherently more quiet.

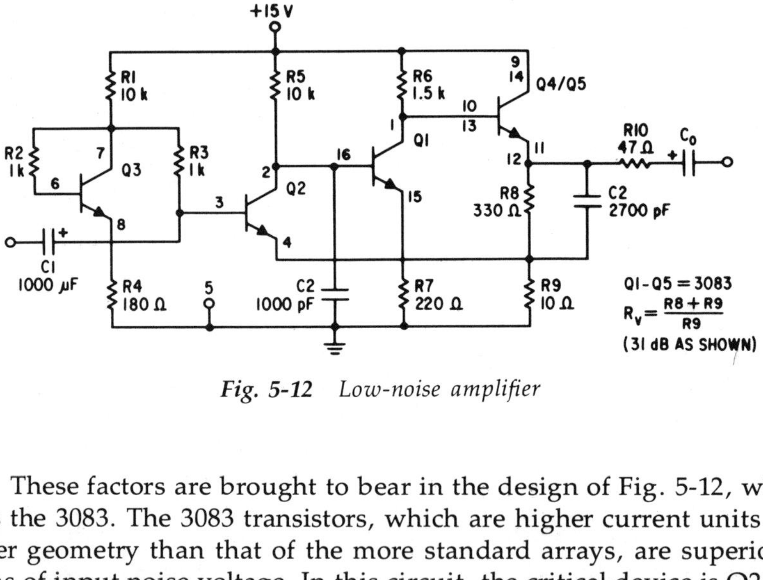

Fig. 5-12 Low-noise amplifier

These factors are brought to bear in the design of Fig. 5-12, which uses the 3083. The 3083 transistors, which are higher current units of a larger geometry than that of the more standard arrays, are superior in terms of input noise voltage. In this circuit, the critical device is Q2, the input stage. Here, it is operated at 1.3 mA to minimize the input voltage noise, and at a high gain to minimize the noise of the second stage. The amplifier is a noninverting feedback type, with feedback provided from the R8–R9 voltage divider. R9 is set at 10Ω to minimize the noise contribution of the feedback loop.

Q2 is biased by Q3, connected as a diode, and with a 1-k base resistor R3. A 1.3-mA reference current is forced in Q3 by R1, and R4 is selected to drop about 250 mV. This 250-mV potential (when the loop is correctly dc biased) will be seen also at the emitter of Q2, via dc feedback from the output. Since the output stage operates at a relatively large current of 25 mA, two transistors are paralleled, Q4 and Q5.

The circuit has a gain of 31 dB, as determined by the quotient when the sum of R8 and R9 is divided by R9. The equivalent input noise is on the order of 1.5 $nV/\sqrt{HZ}$, which, when referred to a 20-kHZ bandwidth, produces an input noise of 0.2 μV (rms). This noise at the output will be 7.5 μV, or 82 dB below a 100-mV output level.

While this is quite good performance in terms of low noise, it should be pointed out that it can only be realized with the value of C1 noted and with a low noise power supply. If C1 is not sufficiently large (at least several hundred μF), low frequency noise will be produced because of its reactance.

The circuit has a maximum output of 3 V rms, and distortion is below 0.03 percent for levels below 1 V of output. Bandwidth is set by C2 at 180 kHz, as shown. The circuit can drive loads of 500 Ω or more. The combination of performance characteristics makes it a good choice for a low-level phono *pre-preamp*, a low-noise, gain-boost device used with moving coil cartridges. It is also suitable, of course, for other low-impedance (100 Ω or less) signal sources with outputs of 250 mV pp or less.

Phono / Microphone Preamplifiers

A more generalized type of feedback amplifier is the phono/microphone preamplifier circuit shown in Fig. 5-13. This circuit can actually be manipulated for a variety of tasks simply by using alternative feedback networks, as appropriate. Thus, it is potentially more general-purpose than its name may imply.

The circuit uses all five transistors of a 3183 high-voltage array and operates from a 35-V power supply. Q1–Q4 operate signalwise as a two-stage feedback amplifier. Q1 is the input stage operated at a 75-μA current for low input-current noise and minimum bias current. Q2–Q3 operates as a Darlington second stage, which minimizes loading on R3 and allows high gain to be realized. First-stage gain is about 200 and second-stage gain, 50. Overall low-frequency gain is therefore about 10,000 (or 80 dB). For audio-frequency use, gain is more accurately specified by the gain bandwidth product, which is 5 MHz. The emitter follower output buffer, Q4, allows this gain to be realized for loads of 10k or more.

One type of circuit for which a very high feedback factor is desirable is an RIAA (Recording Industry Association of America) phone-equalization preamp. The RIAA equalization standard specifies a low frequency (20-Hz) gain of 20 dB greater than that at the reference frequency, 1 kHz. Thus, a 30-dB, 1-kHz gain circuit will have about 50 dB

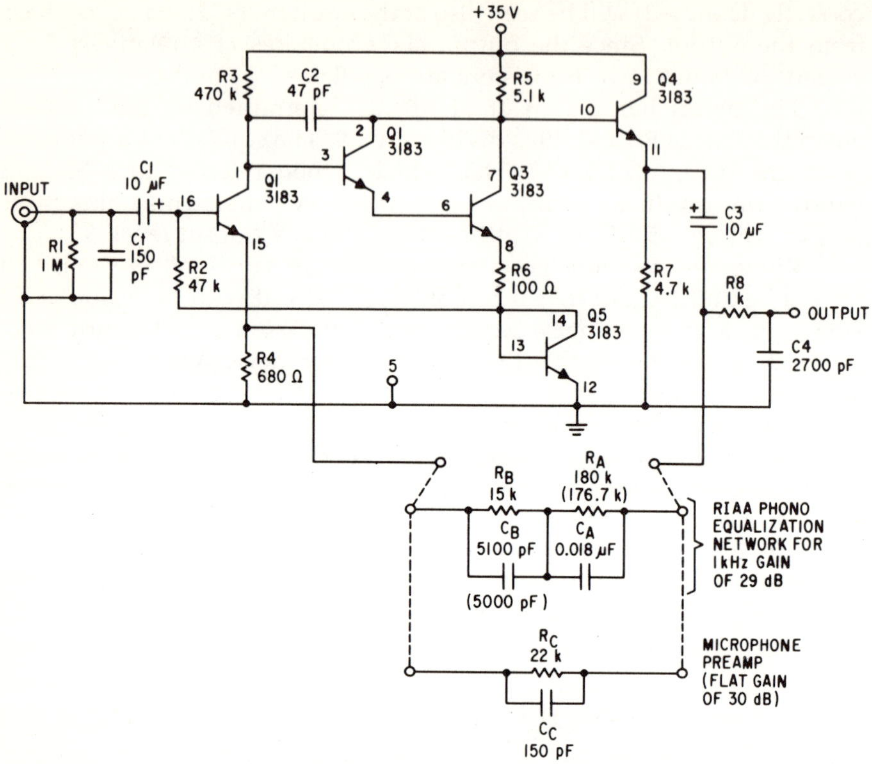

Fig. 5-13 *Phono/microphone preamplifier*

of gain at 20 Hz. If feedback equalization is used, the amplifier *open-loop* gain should exceed this figure, preferably by 30 dB or more for accurate equalization. Thus, an amplifier to be used with an RIAA phono circuit should have a large open-loop, low-frequency gain for good results.

The network shown in the figure for RIAA equalization shapes the response of the preamp in accordance with the required frequency response and uses four standard component values close to the ideal (parentheses).[3] This network, in conjunction with R4, yields a 1-kHz gain of 29 dB. Additional components appropriate to RIAA use are C4 and C_t, the required load capacitance for the cartridge used. The input impedance of the circuit is set at 47k by R2—a value required for phono cartridges almost universally.

The circuit can handle up to 100 mV at 1 kHz and has a 78-dB signal-to-noise ratio with a 10-mV input. Distortion is quite low over the entire range—never greater than 0.05 percent at a 1-V output level.

[3] See Ref. (13) for a detailed design process.

To use the circuit as a microphone preamp with a gain of 30 dB, the alternative feedback network of R_c and C_c is used. This provides a 50-kHz signal bandwidth and an 80-dB signal-to-noise ratio (below a 10-mV input).

Alternative feedback networks can be employed as desired. A variable-gain, general-purpose amplifier can be realized, for example, by using a series network consisting of a 1.5-k resistor and a 50-k pot. This will allow gain to be made continuously variable over a range of 10 to 35 dB.

Wideband "Super-Beta" Amplifier

An example of the 3095 "super-beta" array applied as a wideband amplifier is shown in Fig. 5-14. This circuit, using the 3095 plus a discrete 2N4126 transistor as a level-shift amplifier, has both high gain and wide bandwidth but does not employ overall feedback.

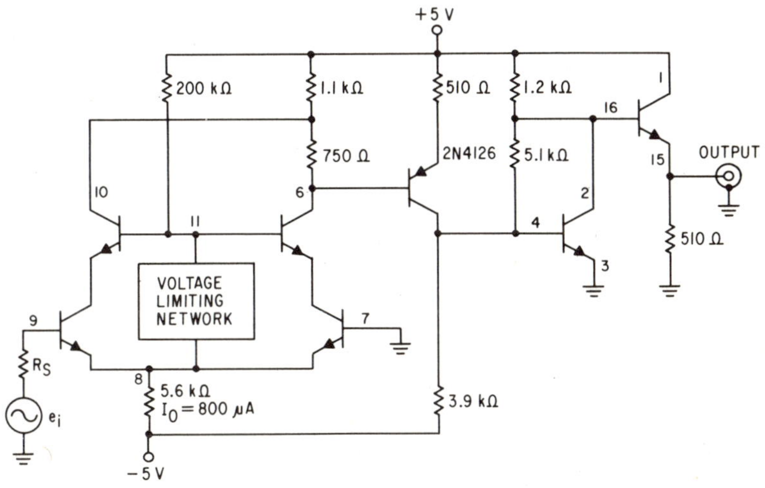

Fig. 5-14 Wideband amplifier using 3095 array (Courtesy, *RCA Solid State Division*)

The 3095 super-beta pair are used as a differential input cascade. A single-ended output is taken from one side, which drives the pnp. A pair of standard transistors from the 3095 are connected as a common-emitter amplifier and emitter-follower buffer, respectively. Overall, the circuit achieves a 33-dB gain and a 3-dB bandwidth of 10 MHz.

Since no feedback is employed around the differential input stage, linearity will deteriorate for higher input levels, or levels beyond about 20 mV pp. As noted, the input circuit is dc-coupled, so that a dc return is necessary. Input-bias current is typically 300 nA (or less) for supply voltages of ±5 V (as noted).

Variable Gain Video Amplifier

Although the bandwidth of IC array amplifier circuitry is naturally quite wide, it can be extended significantly by the use of controlled feedback, where gain is traded for bandwidth. The variable-gain video amplifier of Fig. 5-15 is an example of this design approach and achieves a stable, accurate gain within a very wide bandwidth.

This amplifier consists of a differential input pair, Q1–Q2, loaded in single-ended fashion by R4, with an emitter-follower buffer to drive the load. The circuit form is that of an noninverting op amp, where Q1 is the (+) input and Q2 is the (−) input, or feedback return. In terms of high frequency capability, this configuration is near optimum, as all stages are wideband: Q1 as an emitter follower, Q2 driven as a common-base stage, and Q3 also as an emitter follower. Q3 is operated class A with an 18-mA current to provide sufficient drive current to allow 1 V pp into a 75-Ω output.

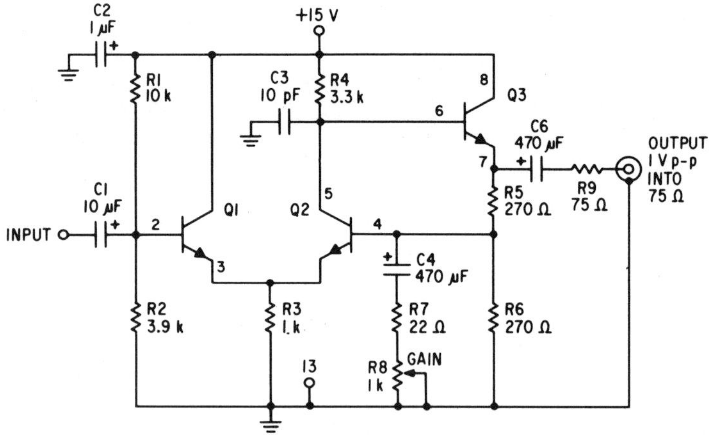

Fig. 5-15 Variable-gain video amplifier

The small signal bandwidth of this circuit is limited more by the natural f_T of the transistors than by the circuit time constants. Large signal bandwidth is limited by the SR (Slew Rate), which is about 100 V/μs. This SR is sufficient to allow sine waves to be reproduced at the output at frequencies up to 15 MHz at the 1-V pp level with a minimum gain of 2. At the higher-gain range extreme of 11, the small signal bandwidth reduces to about 6 MHz and becomes the bandwidth performance limit.

The actual performance realized with the circuit is to some extent dependent upon the layout, particularly the capacitance at the Q2 collector node. C3 may therefore require some adjustment for different physical configurations. Of course, the overall circuit should be compact, with good HF ground and signal path techniques.

The circuit is shown in its widest bandwidth form, and capacitors C4 and C5 have large values from the standpoint of low LF phase shift—a consideration for high-fidelity video response. These can be lessened in size, of course, for modest applications. All capacitors should be low-reactance types, such as tantalums or electrolytics shunted by ceramic discs.

The input impedance of the circuit shown is several kilohms; if this value is excessively low, the unused (Q4) transistor can be employed as an input emitter follower that drives Q1. If even greater output drive is required, Q3 and Q4 can be paralleled; alternatively, a 3083 array can be used.

Video "Line Receiver"

A variety of previous circuits have shown how the matched monolithic properties of IC arrays can be utilized quite effectively in building custom differential amplifiers. These same matched characteristics can be used to advantage in implementing a uniquely useful version of the diff amp, a high common-mode range differential to single-ended amplifier with wideband response and good common mode rejection, termed here a video "line receiver."

In video processing, a common circuit requirement is for a differential input video amplifier that can amplify signals from a coax center conductor to shield but reject common-mode voltages to ground. In this application, input signal levels can be up to 3 V pp of video, with ±5 V of common-mode voltage, which is usually in the form of hum (caused by ground loops). It is also desirable to adjust the video gain with a single noncritical adjustment.

The circuit in Fig. 5-16 is designed around the matched properties of the 3046 array.[4] Q3 and Q4 are a degenerative differential pair with high-impedance dc emitter feeds from Q1 and Q2. With no signal coupling between the Q3–Q4 emitters, the common-mode amplification is very low because of the low output conductance of Q3 and Q4. This results in a common-mode rejection of about 60 dB. The differential signal transfer through R6–R7 results in a unity gain for differential signals, trimmable via R7.

The high common-mode swing capability (10 V pp) is provided by Q1 and Q2, which are biased to a −9-V potential at their bases. Q5

[4] For source, see W. J. Jung, Ref. (11).

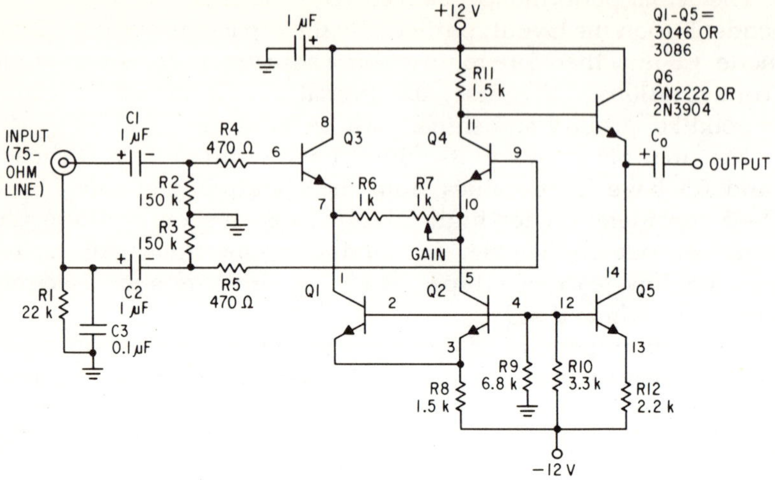

Fig. 5-16 *Video "line receiver"*

regulates the total emitter current of Q6—a discrete output buffer—and Q1–Q2 operate at 1 ma.

The high f_T of the transistors, low circuit time constants, and the common-collector/common-base drive of Q3–Q4 provide the circuit with a video bandwidth in excess of 10 MHz. The input impedance is high enough to bridge a 75-Ω coax line without excessive loading and does not reflect any Miller capacitance to the input line.

A variety of related options can be implemented using this same basic circuit concept. For instance, the 3146 array could be used with ±15-V supplies for an even greater common-mode range—to ±10 V or more. And, of course, the circuit is also useful for an audio line receiver, as described in the next section.

Instrumentation Amplifier

By using the same basic differential input technique shown in Fig. 5-16, a more general-purpose form of the differential input amplifier can be built. This circuit, shown in Fig. 5-17, also has the properties of common-mode noise rejection but is capable of wider ranges of gain and common-mode swing, although within a more narrow bandwidth. The common term for it, an "instrumentation amplifier," will be used here.

In this circuit, Q3 and Q4 are again the differential pair biased by Q1 and Q2, but in this case at a reduced current. Q1 and Q2 are biased at a 75-μA level and at a common-mode limit of −12 V. Q1 and Q2 are loaded by a differential to single-ended converter formed by the R3 to

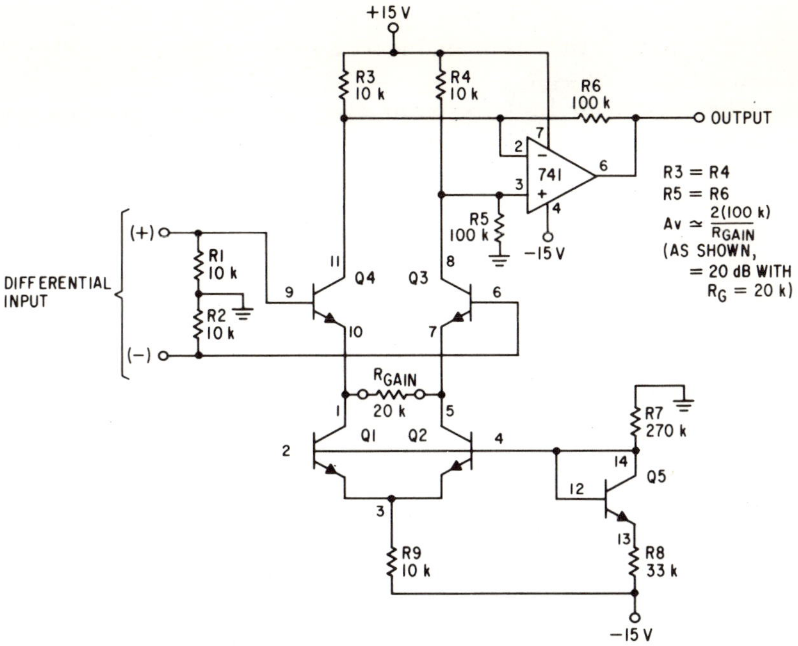

Fig. 5-17 *Instrumentation amplifier*

R6 resistor bridge and the op amp. The bridge resistors are chosen to bias the Q3–Q4 collectors at +12 V. This, in combination with the Q1–Q2 biasing, allows an input range of greater than ±10 V.

Gain of the circuit is set by the ratio of twice R6 to that of the gain programming resistor R_{gain}. For example, with R_{gain} at 20k, the differential input to output gain is 20 dB, or ten times. The R3–R6 resistors should be matched as shown for best common-mode and power-supply rejection.

With the general-purpose 741 op amp, the circuit's bandwidth is 100 kHz, but this may be extended to 300 to 400 kHz if desired by using a TL071 or LF351 (same pinout). Common-mode rejection is best at moderate gains (on the order of 70 dB) and can be as much as 80 dB or more at frequencies below 1 kHz.

Distortion is lowest at higher gains but is usually 0.25 percent or lower for signal levels of ±10 V out. At lower signal levels, the distortion reduces rapidly.

Operational Amplifiers Using IC Arrays

While it is quite obvious that an operational amplifier (op amp) circuit can be fabricated using IC arrays (since IC op amps use simply more of the same technology), what may not be obvious is why one would choose to build his own op amp. For instance, with standard op amp types such as the 741 available for a small fraction of a dollar, this is indeed a logical question.

The reason for building your own is simply to provide specific performance characteristics that are *not* so readily available as those of standard op amps. With the flexibility of IC arrays, one can literally build an op amp optimized around any one specific performance parameter, yet with only a couple of standard array types.

This section discusses several IC op amp circuit design techniques that can be used to satisfy unique performance criteria such as low offset voltage or drift, low bias current, high speed, and so on. The *applications* of operational amplifiers are a subject obviously much broader than can be covered here. For further background on this subject, the reader should consult the author's *IC Op Amp Cookbook.*[5]

General-Purpose Op Amp

To illustrate the basic design procedure, a simple general-purpose op amp is shown as a starting point in Fig. 5-18. This circuit employs a 3096 array as its basis and can be used either directly as shown or modified further, as will be described.

Q1 and Q2 form an npn differential pair that comprises the input stage of this amplifier. Their operating current is set by current source Q3, and they are loaded by the Q4–Q5 current mirror. Q4–Q5 form a differential to single-ended converter that drives the second stage of the op amp—the discrete pnp Darlington, Q6. Q6 has a resistive collector load, R4.

The biasing of Q3 is set up to be within 2 V of the −15-V rail; likewise, Q4–Q5 and Q6 operate within 2 V of the +15-V rail. This gives the amplifier the ability to operate over a common-mode input range of ±10 V (or more) and over an output range of ±10 V (or more).

Overall, this amplifier has a performance similar to that of a general-purpose, standard IC type, and it is unity-gain stable. For the values noted (the *nonparentheses* values, that is), the typical performance will be as shown in the table in Fig. 5-18.

If R6 and R7 are matched to within 1 percent and a 3096*AE* device is used, the input offset voltage will typically be below 1 mV, even without trim. The emitter degeneration of the current mirror and the low base current of the MPS-A65 Darlington second stage provide this low-offset characteristic. If offset trim is desired, R6 and R7 can be modified by including a small percentage adjustment trim, as shown.

The speed of the circuit is largely determined by the operating current of the input stage. With the noted standard values, the unity

[5] See Ref. (12).

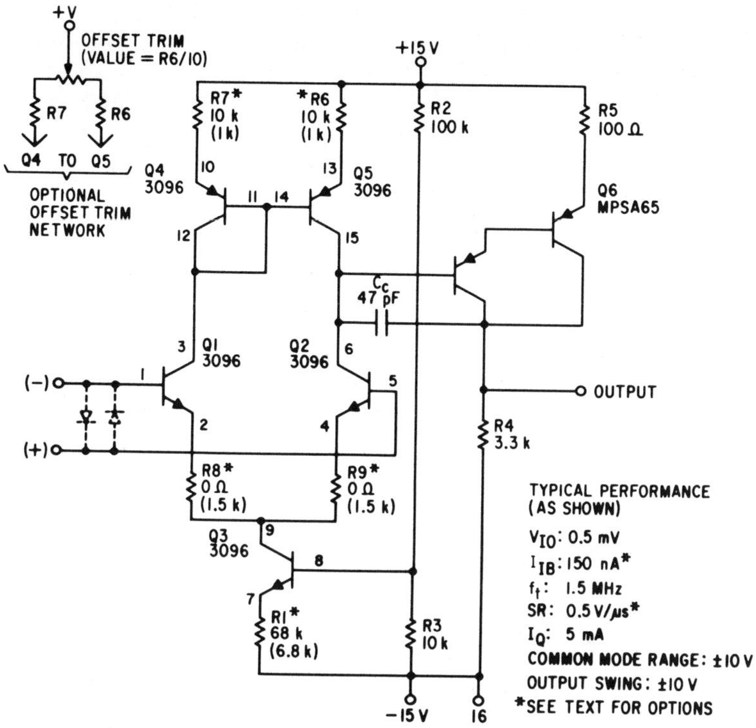

Fig. 5-18 *General-purpose operational amplifier*

gain bandwidth is 1.5 MHz, and the slew rate is 0.5 V/μs. If a higher slew rate (and power bandwidth) is desired, the current in Q1–Q2 can be raised by using lower values for R1, R6, and R7 and by including R8 and R9 in the Q1–Q2 emitter circuit. This will raise the circuit's slew rate to 5 V/μs, but the small signal bandwidth will remain at 1.5 MHz. For either condition, the circuit can be used with any feedback factor desired, down to unity gain. For higher closed-loop gains, such as 20 dB or more, it will be desirable to reduce C_c, the compensation capacitor, proportionately to maintain a high loop bandwidth.

The main limitation of this amplifier is its rather limited output current drive, caused by the simple class A output stage used. For greater current output, R4 can be lowered to about 1k, or alternatively, a class B unity-gain buffer can be added. Even as it stands, however, the circuit is highly useful for lower-level gain stages, such as preamps, where the output drive requirements are modest.

A pair of differential input clamp diodes will protect Q1–Q2 and are strongly recommended to prevent accidents. This precaution applies to *any* npn op amp input stage, whether this one or the others to follow. If a pair of transistors are not available for diode connection from the array in basic use, a pair of discrete units such as IN914 or IN458 should be used.

Composite Amplifiers

A very useful means of increasing op amp performance can be obtained by blending the performance advantages of IC arrays with those inherent in a standard op amp. With careful design, the net composite amplifier can perform on a par with more expensive hybrid or modular amplifiers, yet at modest cost and complexity.

An example of such a composite amplifier is the two-stage op amp of Fig. 5-19. This circuit uses an npn differential pair from a 3146 array as a preamp, ahead of a standard 741 op amp. The combination is then

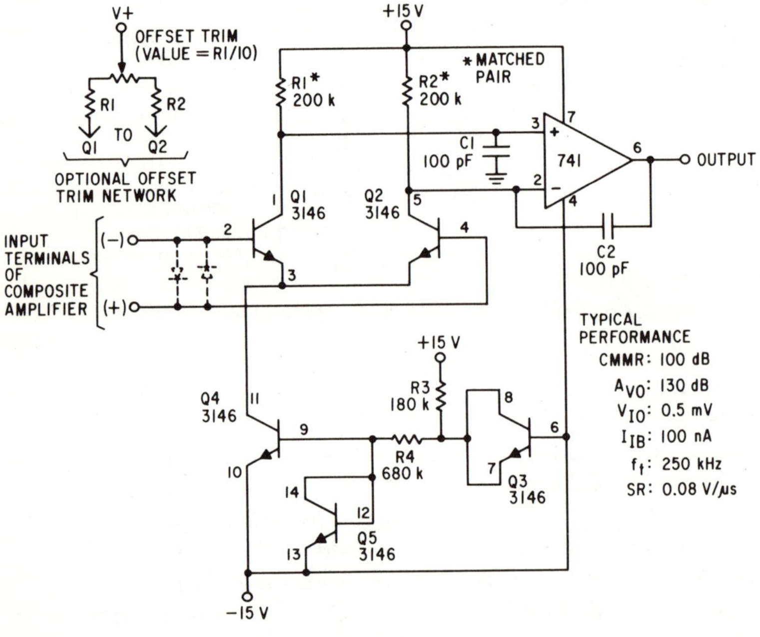

Fig. 5-19 *Composite op amp using IC array as a preamp for the standard op amp*

used as a composite amplifier with overall feedback around *both* stages. The inputs of Q1–Q2 of the 3146 are the two op amp inputs; these are used in normal fashion, with polarity as noted.

A preamp stage such as this, used ahead of the standard IC op amp, can lower the overall noise, offset voltage, and drift and bias current and at the same time increase both the gain and bandwidth. A further substantial improvement is the thermal isolation between the critical input stage and output stage that delivers the load current. The circuit shown is a general example of the technique.

The differential pair Q1–Q2 is loaded by a stable, matched resistor pair, R1–R2. The exact value of these resistors is not overly critical; it should be selected to drop about 1 to 2 V at the operating current of Q1–Q2. Here, Q1 and Q2 operate at 5 μA, and R1–R2 drops 1 V. Note that the op amp actually used for the second stage must be capable of operating with *its* inputs 1 to 2 V below the V+ line; 301As or 741s are adequate in this regard.

Since the Q1–Q2 preamp stage provides additional gain beyond that of the standard op amp, this extra gain must be controlled at high frequencies if the overall amplifier is to be stable when feedback is applied. In this circuit, C1 and C2 provide this function and control both the bandwidth and the slew rate of the circuit in conjunction with the Q1–Q2 operating current. Here, the values provide a bandwidth of 250 kHz and an SR of just under 0.1 V/μs.

The Q1–Q2 emitter current is set up by a Zener preregulator that biases a 9-μA current source composed of Q5 and Q4. This operates Q1–Q2 at 4.5 μA per side, which in turn yields an input bias current of about 100 nA.

As can be noted from the numbers quoted above, speed is *not* a major asset of this amplifier. However, the dc performance is excellent. Offset voltage is typically 0.5 mV (trimmable to zero, using the optional network shown); dc gain is 130 dB or more; and the common-mode rejection ratio (CMRR) is more than 100 dB. Note that these parameters greatly surpass those of the 741, but the good output drive capability of the 741 is still available. This amplifier is well suited to high-gain instrumentation applications or other uses that demand high dc stability.

The circuit as shown is intended to demonstrate the basic design technique and can be optimized even further if desired. For example, use of 3096 transistors for the Q1–Q2 pair will result in bias currents about two to three times lower (50 nA or less) than those of the 3146 units. Another alternative is the use of a Darlington differential pair for Q1–Q2, which can lower bias currents to below 10 nA.

Temperature-Controlled Differential Pair Preamp

Very high performance in terms of dc stability can be achieved by stabilizing the substrate temperature of the preamp section of a composite op amp. Such stabilization holds bias current and offset voltage drifts with temperature to a minimum by the simple expedient of not allowing the temperature of the preamp stage's critical element(s) to change. The technique allows offset voltage drift to be held to below 1 μV/°C.

A circuit that will stabilize a composite op amp such as that just described is shown in Fig. 5-20. Here, Q1–Q2 comprise the stabilized pair. Q3–Q5, in conjunction with the 3140 op amp, provide the temperature regulation for Q1–Q2. This circuit functions similarly to the stabilized Zener reference circuit described in Chap. 4. The op amp used must be one capable of sensing voltages near the V− supply, such as the 3140 shown or a section of a dual 358 unit.

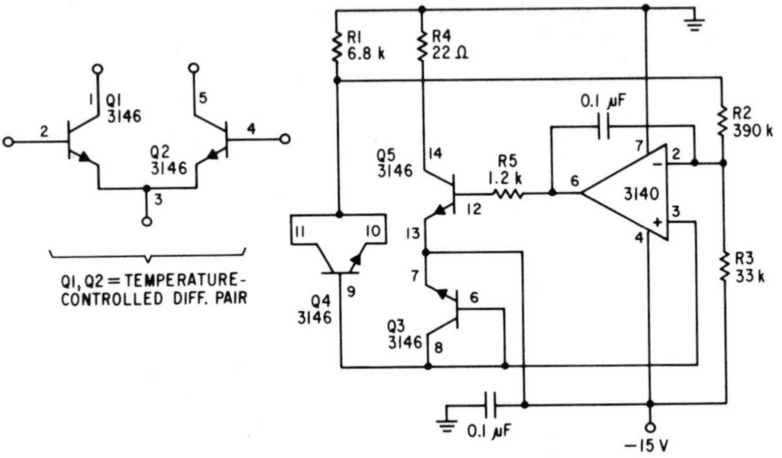

Fig. 5-20 Temperature-controlled differential pair preamp

Super-Beta Input Op Amp

By using the 3095 super-beta IC array for the input stage preamp, it is possible to lower the required input bias current dramatically and thus further improve dc performance capability. A composite op amp circuit that employs the 3095 as a preamp is shown in Fig. 5-21.

This circuit is purposely designed to be similar to the basic configuration of Fig. 5-19 from the standpoint of bandwidth and slew rate to show how the 3095 device, with much higher gain in transistors Q1–Q2, can achieve a lower input bias current with an identical collector current of 4.5 μA.

A logarithmic current source is formed by transistors Q6–Q8. This current source is fed a reference current of 100 μA through R4, and the desired output current of 9 μA is achieved by setting R5 to 6.8k [see the nomograph in Fig. 2-6(b)]. With the current of 4.5 μA per side, the gain of 1,000 or more in the 3095 will result in an input current of 5 nA (typically), more than a order of magnitude lower than that with the 3146 array. It should be noted that to maintain this lower input current,

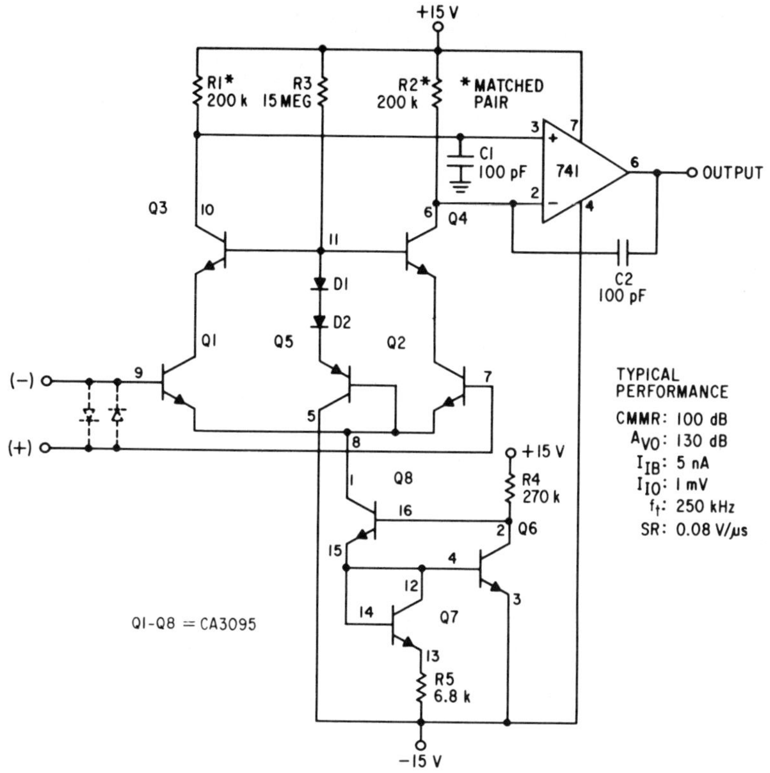

Fig. 5-21 *Super-beta input op amp*

clamp diodes are required so as to prevent Zener breakdown of Q1 and Q2 and the reduction of their gain.

This circuit based on the 3095 will not be capable of as low a voltage drift as the temperature-stabilized 3146 circuit if it is used as shown. The 3095 can be temperature stabilized, however, by connecting Q6–Q8 similarly to Q3–Q5 of the 3146 circuit and using another current source arrangement.

Comparators

Just as op amps can be implemented with IC arrays, so also can comparators. A comparator is, after all, very similar to an op amp, with the exception that it is not frequency-compensated and is therefore capable of useful gains at higher frequencies (higher gain bandwidths).

Some of the reasons for constructing a comparator from an IC array instead of using a standard IC comparator are similar to those for constructing an op amp: more flexibility, the ability to optimize, and so on.

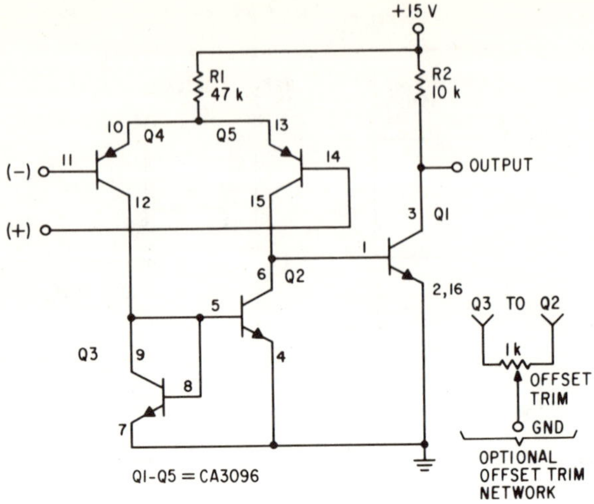

Fig. 5-22 Simple comparator

This section will briefly discuss some practical comparator circuitry using IC array devices.

A very simple comparator employing a single 3096 IC array is shown in Fig. 5-22. Here, Q4–Q5 are a pnp input differential pair that drive the npn current mirror formed by Q3 and Q2. The output of the current mirror and Q5 drives Q1, as a saturated switch. Q1 has a resistive load to V+, R2. Thus, the output-voltage swing of Q1 will be essentially the full supply voltage.

Although the circuit is shown to be operated from +15 V, it may be used at virtually any supply voltage from 5 to 30 V. R1 sets the current in the input stage in the 100- to 200-μA range, which in turn results in a 1-μA (or less) input current.

Since pnp input transistors are used, there is no limitation on differential input voltage as there is with npn devices. Input voltage limits can thus be greater than 7 V (but should, of course, be *less* than the limits of the supply lines).

Improvements possible for the simple circuit are replacing R1 with a fixed-current source or using a Darlington differential pnp pair (with devices from a 3084 array). Offset voltage is not adjustable in the circuit shown, but it can be trimmed to zero, if desired, by adding resistors in the ground legs of Q2 and Q3 and varying their relative ratio (see optional network).

General-Purpose Comparator

Figure 5-23 illustrates a more general-purpose comparator, one less restricted in terms of its input common-mode voltage range. It uses an npn input stage biased by a current source and can therefore compare small differential voltages over a common-mode range of ±10 V (or more) with no substantial change in bias currents.

The circuit can be considered a mirror image of the simple comparator, with the exception of the current source transistor, Q3. A 3096 array is used again, with a 2N3906 discrete transistor as the saturated output switch. With bipolar supplies as shown, the output will swing from (+) to (−) 15 V.

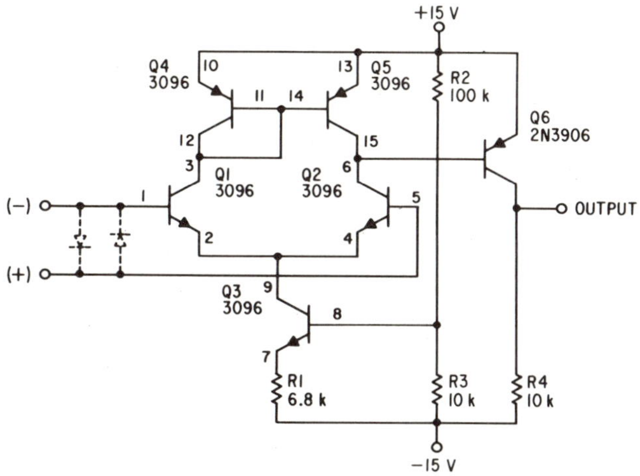

Fig. 5-23 General-purpose comparator

REFERENCES FOR CHAPTER 5

(1) Mortensen, H. H., "A Fully Differential Input Voltage Amplifier," National Semiconductor LB-20, Dec. 1972.

(2) Dobkin, R. C., "IC Preamp Challenges Choppers on Drift," National Semiconductor AN-79, Feb. 1973.

(3) Dobkin, R. C., "Instrumentation Amplifiers," National Semiconductor LB21, June 1973.

(4) Lovelock, P. A., "Integrated Circuit Arrays," *Ham Radio*, Sept. 1978.

(5) Kleinman, H. M., "Design Ideas for RCA Linear Arrays," RCA ST-3895, Dec. 1968.

(6) Jung, W. G., "A Survey of Application Ideas for IC's," *Broadcast Engineering* (three part series), Sept., Oct., Dec. 1969.

(7) Bray, D., "Audio Preamplifiers," Monochip Application Note APN-11.

(8) MacPherson, E., "Integrated High Speed Operational Amplifier," Monochip Application Note APN-7.

(9) Grebene, A. B., *Analog Integrated Circuit Design,* Van Nostrand Reinhold, 1972.

(10) Jung, W. G., "Monolithic-Triad Current Summer," *EDN/EEE,* July 1, 1971.

(11) Jung, W. G., "10 MHZ Differential Video Line Receiver Built with Monolithic IC Transistor Array," *Electronic Design,* April 26, 1974.

(12) Jung, W. G., *IC Op Amp Cookbook,* Howard W. Sams, 1974.

(13) Lipshitz, S. P., "On RIAA Equalization Networks," presented at 61st AES Convention, Nov. 1978 (preprint # 1424).

(14) ______, "101 Analog IC Designs," Interdesign, Inc., 1976.

(15) ______, "Linear Integrated Circuits," RCA Technical Series IC-42, 1970.

6
Oscillators

A very common circuit requirement is an oscillator, or signal source that will produce continuous signals of various waveshapes. Typical of these are clock sources, pulse generators, waveform generators, sawtooth, ramp, or triangle generators, and the familiar sine-wave signal source.

The flexibility of IC array circuit techniques allows relatively easy and convenient implementation of circuits of this type. A variety of them will be discussed in this chapter. As was the case in many of the previous circuit examples, the circuits are directly useful as shown, but most of them are also readily adaptable to other needs as well.

Comparator Astable

One of the simplest, easily designed astable oscillators is shown in Fig. 6-1. This circuit, called simply a "comparator astable," generates square or rectangular waves over a wide range of frequencies.

The circuit is called a "comparator astable" because its operation is based on a voltage comparator (or op amp) that is operated with positive feedback. The positive feedback network determines the switching thresholds of the oscillator, whereas a negative feedback network determines an RC time constant and thus the frequency of oscillation. In the circuit of Fig. 6-1, the 3096 transistor array and its associated components form the comparator; the R_t–C_t and R3–R5 are components associated with the oscillator function and determine frequency.

In the comparator, Q4 and Q5 operate as a pnp input differential amplifier that allows a high differential input rating. This factor removes restrictions on supply voltage that would arise as a result of V_{EB} breakdown if an NPN input stage were used.

Q4 and Q5 are operated at approximately 100 μA of emitter current, which in turn produces a 1.5-μA base current. This allows the use of timing resistances of up to 100k, with small errors caused by offset.

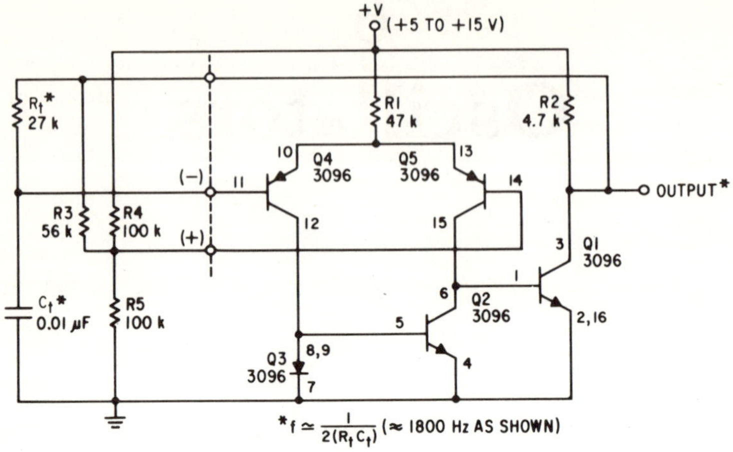

Fig. 6-1 *Comparator astable*

Moreover, the 100-μA operating current provides reasonably fast switching times by providing a relatively high drive to Q3–Q2 and Q1.

Q3 and Q2 form a simple current mirror that provides a push-pull drive to Q1. Q1 is operated in a switching mode, either saturated or cut off. Thus the output of the comparator is a square wave that swings from ground to the V+ level.

The circuit operates over a fairly wide range of supply voltage—+5 to +15 V—with relatively small change in frequency caused by supply voltage. It is somewhat sensitive to loading, however, because of the simple resistive pullup load (R2) for Q1. If a buffer stage is added, this effect can be easily eliminated.

If R3–R5 have the values shown (or the same ratios if other values are substituted), the frequency of oscillation is expressed by the simple equation shown. For best stability in this and other RC oscillators, the timing components should be stable types, for example, carbon film resistors and film type capacitors such as polystyrene or polycarbonate capacitors. Polarized electrolytic types can be used (with the negative terminal grounded) if their leakage is low—preferably 1 μA, or less.

Since this is basically such a useful circuit, many different types of modifications are possible. For simple variation of frequency, R_t can be a rheostat. If it is desired to trim the duty cycle, R_t can be replaced by two resistances, each in series with a diode (with the two diodes reversed in polarity). Alteration of the relative values of these two resistances will then adjust the output duty cycle.

If low-frequency operation is desired, a low-power version of the circuit can be realized simply by scaling resistances R1–R5 upward by a factor of ten. This will allow R_t to be 1 MΩ or more, and the entire circuit will draw less than 100 μA (at 5 V).

Emitter-Coupled Astables

Since IC array devices lend themselves extremely well to the emitter-coupled variety of oscillator, a number of them will be discussed in this section. Emitter-coupled astable oscillators are useful from ultra-low frequencies well into the megahertz range. They are relatively easy to design, fairly simple, and offer good performance with regard to stability.

Low-Frequency Emitter-Coupled Astable

A general-purpose, low-frequency-range version of an emitter-coupled astable is shown in Fig. 6-2. This is a fairly simple three-transistor circuit and can employ either a 3086 or 3046.

In operation, Q3 and Q4 form the emitter-coupled oscillator pair. Q3 is biased at a fixed-base potential by R1–R2 and D1. The regenerative feedback path is from the collector of Q3 to the base and emitter of Q4, then through C_t to Q3 as a common-base amplifier. Q1 functions only as a buffer amplifier and isolates the signal at the collector of Q3 from load variation.

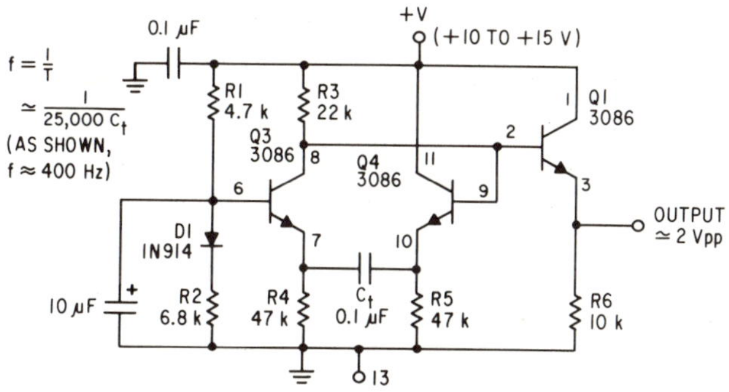

Fig. 6-2 Low-frequency emitter-coupled astable

A number of circuit values determine frequency in this circuit, namely R3, R4, R5, and C_t. For simplicity of use, R3, R4, and R5 are fixed in value here; as a result, C_t can be simply changed for different frequencies by using the equation shown.

The circuit values shown produce an output at a frequency of 400 Hz, but the circuit is useable from as low as 10 Hz to as high as 100 kHz.

A continuous or smooth frequency trim can be accomplished by a small change—making R3 a pot plus fixed resistance. Doing so also produce changes in output amplitude, however, and these may be undesirable. The circuit produces the 2-V pp output shown, nominally a square wave. If a trim for symmetry is desired, the ratio of R4 and R5 can be adjusted.

Options for this circuit can go beyond the minor alterations mentioned. A very low-power version can be implemented by scaling resistances upward by a decade. The circuit is fairly independent of supply voltage over the range shown.

Medium-Frequency Emitter-Coupled Astable

A modified version of this emitter-coupled oscillator is shown in Fig. 6-3. It is called a "medium-frequency emitter-coupled astable" because it extends the usable frequency range upwards to about 1 MHz.

One of the reasons for the confined speed range in the circuit of Fig. 6-2 is that Q3 of the oscillator loop saturates for a portion of the cycle, necessarily inhibiting speed. In Fig. 6-3, Q3 operates as before, but the relative circuit impedances are lowered and higher current transistors are used from the 3083. This technique allows the oscillator to perform to 1 MHz or more with reasonably good predictability.

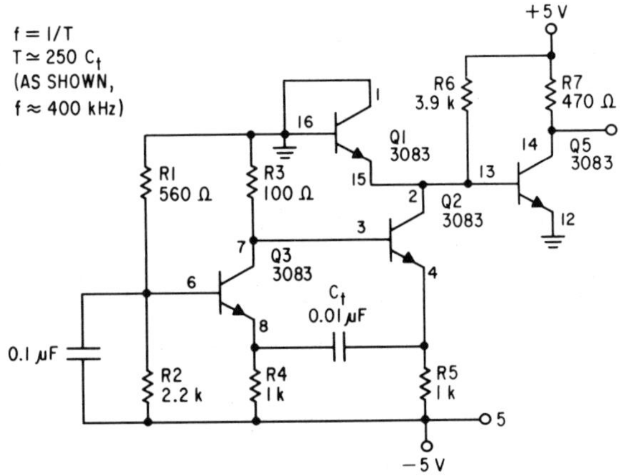

Fig. 6-3 Medium-frequency emitter-coupled astable

In this circuit, Q3 and Q2 comprise the oscillator, which operates from a −5-V supply. This feature allows the collector of Q2 to drive the base of Q5 directly, which in turn provides a saturated 5-V output that is suitable for TTL uses. Q1 acts as a clamp for Q5 to prevent Zener action.

The circuit as shown oscillates at about 400 kHz, and Ct is simply changed for other frequencies. R3, R4, and R5 perform functions similar to those of their counterparts in Fig. 6-2.

General-Purpose Emitter-Coupled Astable

A slightly different, but highly flexible, version of the emitter-coupled oscillator is shown in Fig. 6-4. In this configuration, Q3 and Q4

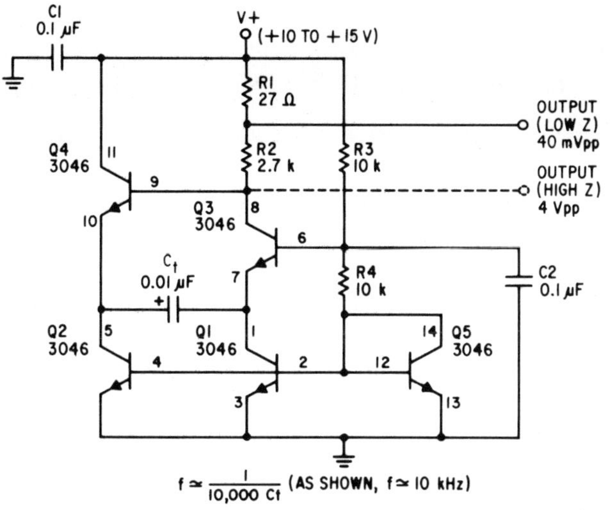

Fig. 6-4 *General-purpose emitter-coupled astable*

are again emitter-coupled by a timing capacitor, C_t, a main difference in this case being that the emitter currents of Q3 and Q4 are determined by constant-current sources Q1 and Q2. The constant-current biasing of Q3 and Q4 prevents the saturation of Q3 during the timing cycle, resulting in a number of advantages.

The bias for the entire circuit is set up by the R3–R4 voltage divider and diode-connected Q5; in turn, Q1 and Q2 are biased at the same current as that flowing in R3–R4. The R3–R4 junction, bypassed by C2, biases the base of Q3, which functions as a common-base amplifier.

Timing capacitor C_t is charged (and discharged) with a current equal to that in Q5, and thus in Q1 and Q2. The current through Q3 alternates between twice this current and zero, thus forming a symmetric square wave at the collector of Q3. For the values shown, this square wave is 4 V pp in level (for a 15-V supply).

An interesting feature of this circuit is that the output voltage varies in a similar proportion to the timing currents with different supply voltages. As a result, operation is only midly sensitive to supply voltage, with less than a 1-percent frequency change for a 1-V supply change.

The constant current biasing of the circuit ensures that neither oscillator transistor is allowed to saturate during the timing cycle, thus preventing loss of gain and the storage delays associated with saturation and allowing fairly high-speed operation, even with relatively low currents. The frequency expression is simplified, as shown in the figure. Operation is quite predictable up to about 1 MHz, where stray capacity begins to introduce errors. For best results in terms of accuracy, the 4-V output should not be used directly because of its sensitivity to capacitive loading. A simple emitter-follower stage can be used to minimize this error source.

An interesting side feature of the circuit is the fact that the emitters of Q3 and Q4 are offset dc-wise, which means that C_t can be a polarized type. Best results at low frequency will be realized with tantalum electrolytics because of their closer tolerances. Due to this consideration and the high frequency capability mentioned, this oscillator is useful just as shown from a range of below 1 Hz to above 1 MHz. Either a 3046 or 3086 array type can be used, or the oscillator can be repinned for another low-voltage array type. Operation at higher supply voltages is not recommended, because Q3 and Q4 must be protected against emitter-base Zener breakdown. At 15 V, they see a maximum of 4-V reverse bias, which is safely conservative.

Possible modifications to this design might be for very low-frequency or very high-frequency operation as a means of extending its range at either end of that quoted. Note that it will probably not be possible to push performance capability in *both* directions simultaneously, simply because optimizing the circuit for high speed (for example) will sacrifice LF range (and vice versa).

High-Speed Emitter-Coupled Astable

The emitter-coupled astable circuit of Fig. 6-5 is a very high-speed version of the comparator astable that uses a nonsaturating voltage comparator. The comparator is formed by the emitter coupled-pair Q1 and Q2 from a 3086 array.

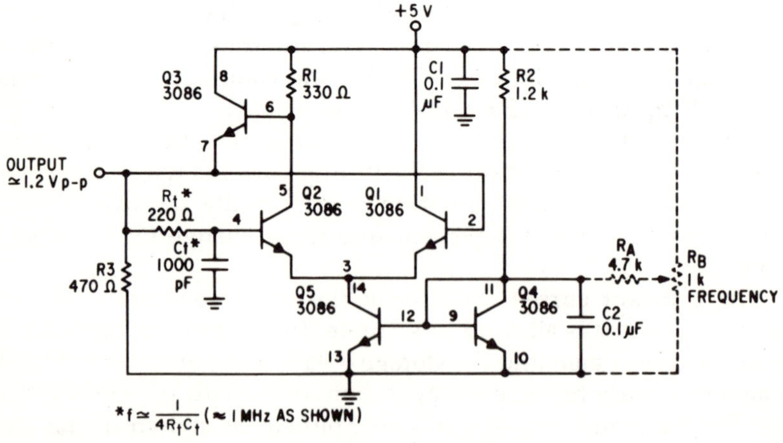

Fig. 6-5 *High-speed emitter-coupled astable*

In this circuit, current source Q5 is biased at a 3.6-mA level by diode-connected Q4. This constant 3.6-mA emitter current to Q1–Q2 will be toggled to either Q2 or Q1, depending upon their relative differential input voltage. When Q2 is fully on, a 1.2-V drop is created across R1. This drop determines the pp output level of the square wave, which is buffered by emitter follower Q3.

The positive feedback of the astable is direct-coupled to the base of Q1, whereas the negative feedback path leads to Q2, via R_t and C_t.

The frequency of oscillation is determined by the expression shown. Best predictability is achieved for R_t values of 100 Ω to 1k Ω, and C_t values of 20 pF or up. Because of the high frequency operation of which the circuit is capable, a compact, minimum-capacitance layout is desirable. The circuit is useful to frequencies above 10 MHz.

Frequency stability versus supply-voltage changes is not one of the virtues of the circuit, and supply voltage should be regulated for best stability. As a matter of fact, frequency varies in proportion to the voltage, a factor that can be used to make the circuit voltage tuned.

If an (optional) current-summing resistor such as R_A is connected to Q4 as shown, a variable voltage applied to this resistor can be made to tune the oscillator frequency. As shown, it is manually tunable via pot R_B, but an AFC (or other) control voltage could be applied to R_A as well.

Stabilized Emitter-Coupled Oscillator

While the emitter-coupled astable just described is useful at high frequencies and attractive for its simplicity, it is sensitive to supply-voltage changes. When a stabilized frequency oscillator circuit is required, this astable can be modified rather easily.

A stabilized emitter-coupled oscillator is shown in Fig. 6-6. It is identical to the circuit of Fig. 6-5 except for the frequency-determining components. In this circuit, the frequency is determined by a crystal from Q2 to ground.

The crystal provides an effective low impedance at its resonant frequency. Since it is in the negative feedback path and in shunt, this provides negative feedback to Q_2, at all but the crystal frequency. The circuit therefore sees greater positive than negative feedback at this frequency, at which it oscillates.

With a 1-MHz crystal, the supply sensitivity is low—less than 20 ppM/V, typically. An alternative form of this circuit can be achieved by using an LC tank in place of the crystal, as shown. Supply sensitivity of this form of the circuit is about 0.5 percent/V. Although this value is relatively poor compared to that of the crystal circuit, it is still far superior to that of the RC astable circuit.

An added virtue of both the crystal and LC stabilized circuits, of

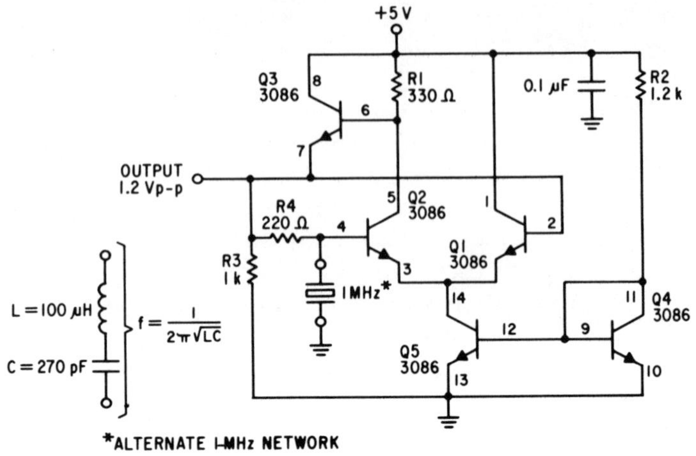

Fig. 6-6 *Stabilized emitter-coupled oscillator*

course, is their good frequency predictability, which can be quite high in the case of the crystal (to tolerances of 0.01 percent or better).

General-Purpose Waveform Generator

When ramp, sawtooth, or triangular waveforms are required, the waveform generator circuit of Fig. 6-7 can be quite useful. This circuit produces both linearly rising and falling voltage ramps as well as rectangular pulses. Timing periods are completely adjustable and independent of one another, without any limit on the duty factor.

The circuit is based on a 555 timer IC, which offers the desirable capability of integral high- and low-threshold voltage comparators as well as a high-level square-wave output. A 3096 pnp/npn IC array is used as a controlled source/sink of current that charges and discharges timing capacitor C_t.

During timing period t1 (see waveform diagram), the 555 output is high, and diode D1 allows Q5 of the array to charge C_t toward V+. Simultaneously, pin 7 of the 555 (an open-collector output) is in the off state, preventing Q1–Q2 from conducting. When the voltage across C_t reaches the upper 555 threshold of $\frac{2}{3}$ V+, it changes state, beginning timing period t2.

During period t2, the 555 output is low, and pin 7 and Q1–Q2 conduct. D1 steers the current away from Q5, removing the charging current from C_t. The current from Q4 is mirrored by Q1–Q2 and discharges C_t toward ground. The C_t voltage decreases until it reaches the lower 555 threshold of $\frac{1}{3}$ V+, whereupon time period t2 ends and the cycle repeats with t1 again.

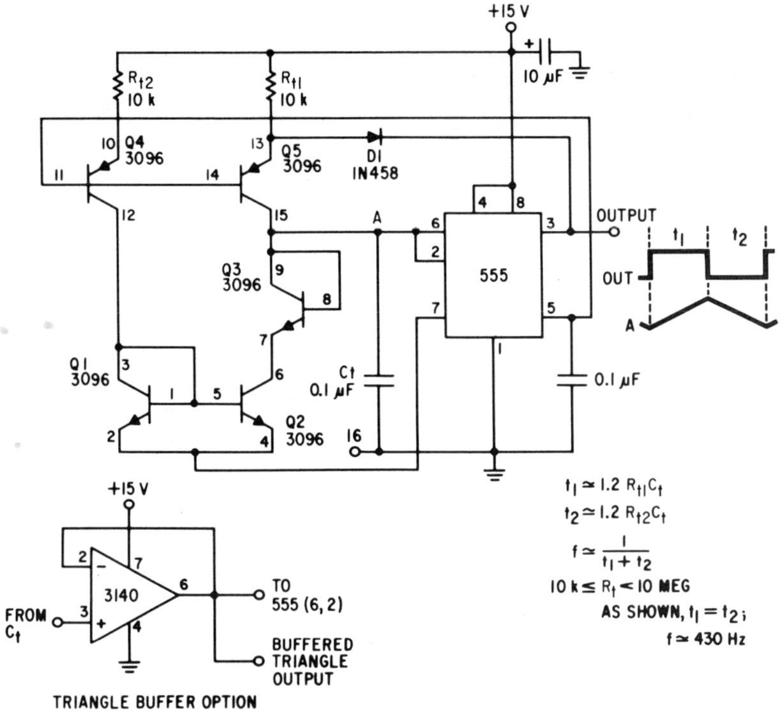

Fig. 6-7 General-purpose waveform generator

Time periods t1 and t2, which are expressed by the simple relationships shown, are not strongly sensitive to supply-voltage changes. For a given value of C_t, the ratio of R_{t1} to R_{t2} controls the duty cycle, and the range of values for R_t allows duty factors as high as 1000/1 (or 1/1000). With the values shown, the duty cycle is nominally 50 percent and the frequency, 430 Hz.

If the triangle or ramp output across C_t is desired to be used externally (as will normally be the case), a buffer amplifier between C_t and the 555 will be helpful. A simple solution is to connect a 3140 op amp as shown.

The circuit is usable from a frequency well below 1 Hz to as high as 100 kHz. Note that if suitable controls are provided for R_{t_1}, R_{t_2}, and C_t, a wide range of waveforms can be implemented rather inexpensively.

Triangle-to-Sine-Wave Converter

A very useful companion circuit to the waveform generator of Fig. 6-7 is the triangle-to-sine-wave converter of Fig. 6-8. This circuit—a form of nonreactive filter—is capable of converting a constant-amplitude traingle wave to a low-distortion sine wave.

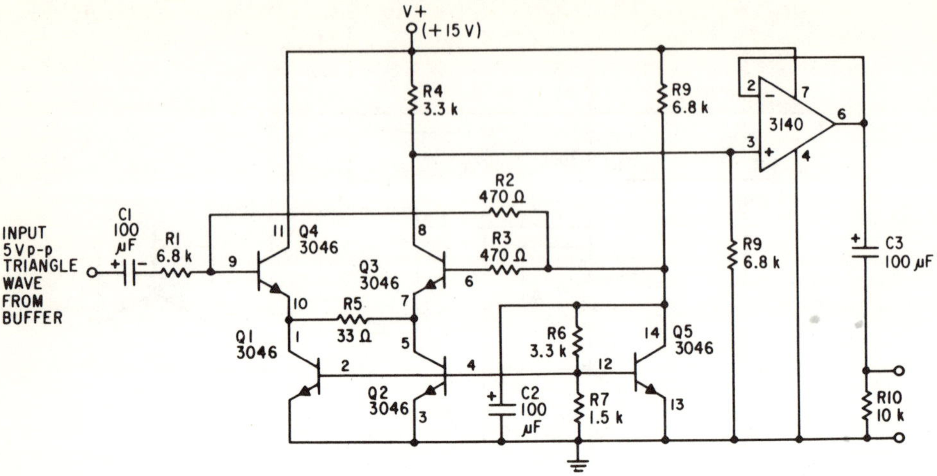

Fig. 6-8 *Triangle-to-sine-wave converter*

For a constant-amplitude traingle wave input, the circuit produces a constant-amplitude sine wave output over a wide frequency range.

This circuit operates upon the principle of nonlinear gain in a differential pair. Q4 and Q3 are a matched pair from a 3046 and fed by currents sources Q1–Q2. R5 provides controlled emitter coupling between Q4 and Q3. With a fixed differential input to Q4 of about 350 mV, R5 can be optimized to produce a minimum of distortion in the output sine wave from Q3.

For the values shown, the circuit has a unity "gain" between a 5-V-pp triangle and a 5-V-pp sine wave at the output. The input is ac-coupled through C1, which is fed from the circuit of Fig. 6-7 (with buffer option included).

Typical performance realizable with this technique is a sine-wave THD (Total Harmonic Distortion) of 1 percent or less. This waveform fidelity is illustrated by the actual scope photo of the input and output signals in Fig. 6-9.

The output from Q3 is a high impedance, about 2.2k for the values shown. A unity-gain op amp buffer is therefore recommended, one similar to that shown in Fig. 6-7. Note that this can be a dual op amp such as the 3240 (appropriately repinned).

High-Frequency Sine-Wave Oscillator

For applications requiring a single-frequency sine wave with its low harmonic content, a form of linearly operated oscillator is needed. An example of this type of oscillator, optimized for RF or general HF use, is shown in Fig. 6-10.

This circuit uses two 3086-type arrays, one in the oscillator circuit (ICA) and the second in the bias AGC loop (ICB). In the oscillator, Q1-A and Q2-A form an emitter-coupled pair, one side of which is loaded by a resonant tank—L1 and C1–C2. The base of Q2-A is ac-grounded by

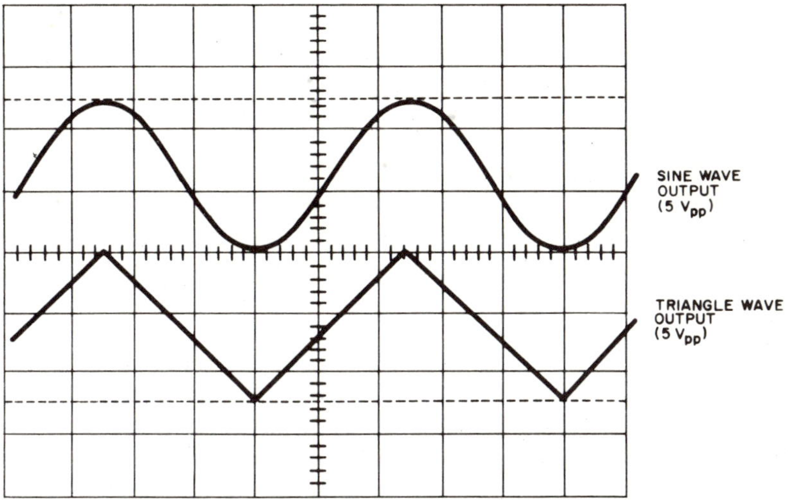

Fig. 6-9 Input-output waveforms of triangle-to-sine-wave converter

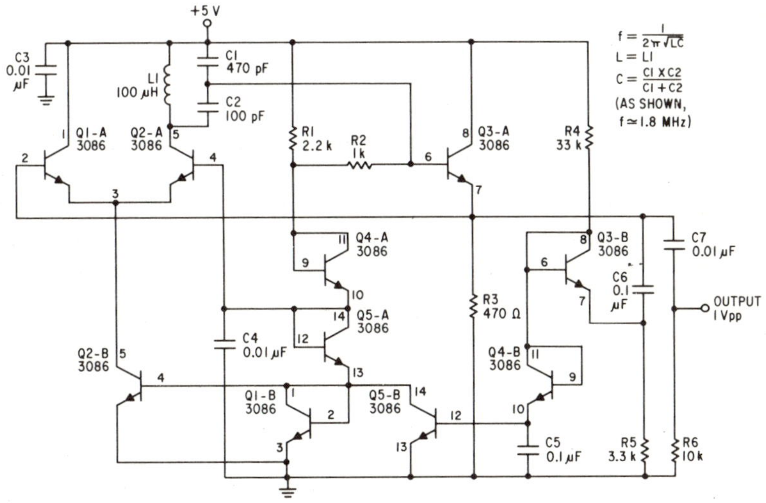

Fig. 6-10 High-frequency sine-wave oscillator

the bias diodes and C4. Therefore, a 1-V-pp since wave fed to Q1-A is effectively a 1-V-pp differential input to the emitter-coupled pair.

A 1-V-pp drive to a differential pair is sufficient to overdrive it into current mode limiting. For such conditions, the output from Q2-A consists of a current square wave. This current, when fed into the tank circuit, produces a sine wave at the tank's resonant frequency.

Because the tank circuit is loaded lightly, a high peak-peak sine-wave ac swing will tend to be developed across it. A portion of this swing is tapped off by capacitive divider C2–C1 and buffered by Q3-A. The output from Q3-A is the output waveform of the oscillator, which is ac-coupled to R6 by C7.

To maintain the tank voltage at a level below the breakdown voltage of Q2-A and its clipping level, some form of AGC is necessary. This is provided by Q2-B, which is a variable source of current for Q1-A and Q2-A. Control of their emitter current controls their effective transconductance and thus their output current, in turn controlling the sine-wave amplitude developed.

A portion of the oscillator's ac output is peak-rectified by Q3-B and Q4-B and stored in capacitor C5. The voltage across C5 is the V_{BE} of Q5-B, which controls the V_{BE} of Q2-B and thus the current fed to Q1-A and Q2-A. This AGC loop holds the output at a steady 1-V-pp level, forcing the LC tank to an operating point at which it will minimize harmonic generation.

The frequency of operation is determined as shown in the figure; it is not limited to the 1.8-MHz example shown. The circuit is useful to well above 10 MHz, with suitable high-Q tank components. For correct AGC loop operation, the ratio of C1–C2 should be maintained close to that shown.

Low-Frequency Sine-Wave Oscillator

For low-distortion sine waves at audio frequencies, an RC oscillator based on the Wien bridge or other filter technique is often called upon. This type of oscillator, like that just described, has a basic design problem in that it must be AGC'd for optimum output-waveform linearity. One form of AGC useful at medium frequencies is the diode clipper; it can employ either the forward or reverse junction characteristic so readily available in IC arrays.

A Wien-bridge oscillator with an interesting form of amplitude stabilization is shown in Fig. 6-11. The Wien oscillator, or variations of it, is often seen in circuit literature, and the basic requirements for frequency determination are listed in the notes. When the network components have the ratio shown in Fig. 6-11, the op amp circuit must furnish a gain of 3 or more (to make up the network loss) for oscillations to start.

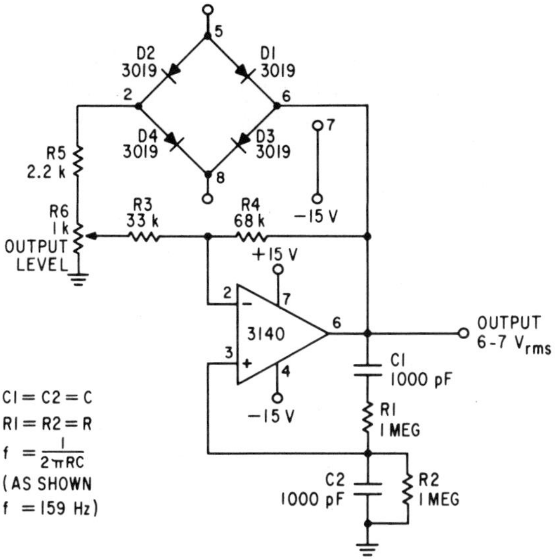

Fig. 6-11 *Low-frequency sine-wave oscillator*

In essence, this means that the negative feedback network must operate around a gain of 3 at the output level desired. Such operation implies some form of AGC, for the gain must exceed 3 for oscillation to start but must be reduced and maintained at 3 for stable, undistorted sine waves to be produced.

In the circuit shown, a 3019 IC array is used to shunt R4 on (+) and (−) waveform peaks, thus, in effect, reducing the instantanous value of R4. Care must be taken that this clipping action not be excessively abrupt or asymmetrical, for excessive distortion will be generated.

Symmetry of clipping is provided by the matching nature of the IC, which provides matched Zener breakdown levels for positive and negative swings. Control over the degree of clipping is provided by R6, which adjusts the output level. With these safeguards, the THD is held to a level of approximately 1.5 percent at output levels of 6 to 7 V rms. Lower distortion is possible at lower levels, but the level must not be set too low or oscillation will not start reliably.

Another factor somewhat critical to the circuit's ultimate performance is the matching and tolerance of network components. For the network loss to be 3 as described and the frequency to be as shown, the R and C elements must be matched. While the circuit will oscillate with reasonably loose tolerances in R1, R2, C1, and C2, lowest distortion and best accuracy is realized with matched, close-tolerance components.

Figure 6-12 illustrates the performance possible when these conditions are met—a 6-V-rms sine-wave output with a THD of about 0.7 percent. The Wien components were within less than 1 percent of the values shown and, as a result, so was the frequency.

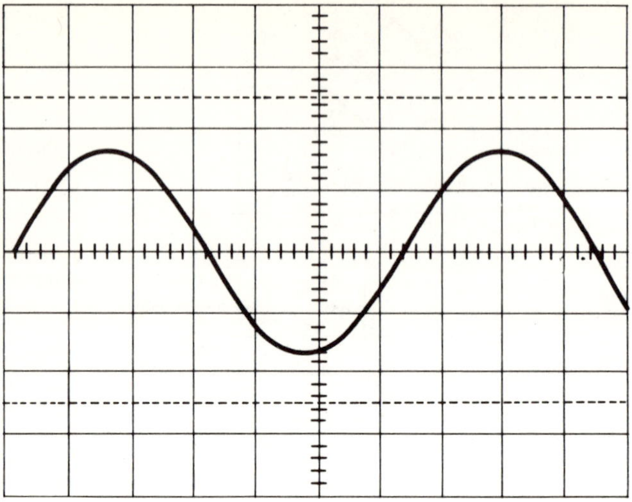

Fig. 6-12 Output waveform of low-frequency sine-wave oscillator

REFERENCES FOR CHAPTER 6

(1) Beneteau, P. J., and Evangelisti, A., "An Improved Emitter-Coupled Multivibrator," Fairchild Application Note 59, Feb. 1963.

(2) Middlebrook, R. D., and Richer, F., "Non-Reactive Filter Converts Triangle Waves to Sines," *Electronics,* March 8, 1965.

(3) Grebene, A. B., "Monolithic Waveform Generation," *IEEE Spectrum,* April 1972.

(4) Jung, W. G., "LED's Do Dual Duty in Sine Wave Oscillator," *EDN,* Aug. 20, 1976.

(5) Jung, W. G., "Sine Wave Oscillators," *db,* July 1976.

(6) Jung, W. G., "Function Generators," *db,* Dec. 1975.

(7) Sheingold, D., (ed.), "Non-Linear Circuits Handbook," Analog Devices, Norwood, Ma., 1974.

(8) Jung, W. G., *IC Op Amp Cookbook,* Howard W. Sams, Indianapolis, Ind., 1974.

(9) Palm, W. A., "ECL IC Oscillates From 10 to 50 MHZ," *Electronics,* May 29, 1975.

(10) Hornak, T., "ECL Tuned Oscillators Are Voltage-Stable," *Electronics.*

(11) Jung, W. G., *IC Timer Cookbook*, Howard W. Sams, Indianpolis, Ind. 1977.

(12) Jung, W. G., *IC Converter Cookbook*, Howard W. Sams, Indianapolis, Ind., 1978.

(13) Data Sheets:

Analog Devices Norwood Ma.	AD537 V/F Converter
Exar Sunnyvale, Ca.	XR2206 waveform XR2207 generator XR2209 IC's
RCA Somerville, N.J.	CA3140 op amp
Signetics Sunnyvale, Calif.	NE555 timer

7
Logic Stages

The general utility and flexibility of IC array devices can often offer unique and efficient solutions to the requirements of nonstandard logic stages. With the overwhelming dominance of TTL and CMOS as the most popular standard forms of logic, no one would reasonably suggest an IC array for efficient or economical *routine* use. Yet, there always exist those "odd ball" requirements that do not justify use of a TTL or CMOS device—for example, special interfacing or power-supply considerations. For these "random" logic situations, the ability to configure an IC array into a variety of logic forms can be an advantage since it eliminates the need for a special IC type.

This chapter covers the use of arrays in such logic stages, most of which can readily be adapted to suit specialized needs. A variety of standard logic functions are possible with arrays, most of which can be realized with a single device.

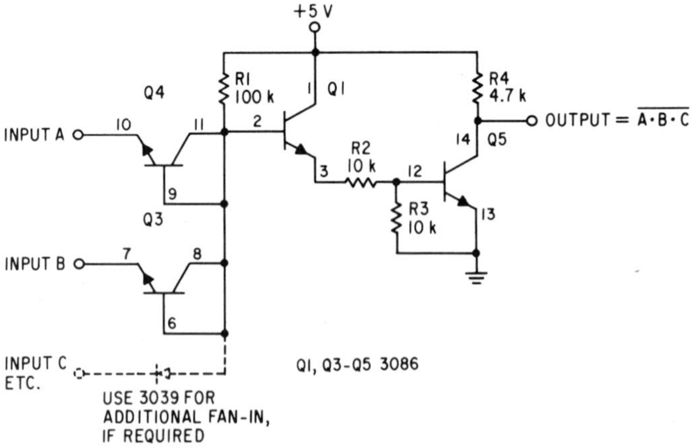

Fig. 7-1 *NAND gate*

Gates

The standard logic building block is the gate, all types of which can be implemented with arrays.

In Fig. 7-1 is shown a NAND gate, which is based on the basic DTL logic structure. Q4 and Q3 are connected as diodes and provide the AND function for inputs A and B. Q1 and R2–R3 provide the functions of buffering and level shifting, and Q5 is an inverter to restore saturated levels. The overall function is NAND, resulting from the AND-INVERT process.

A 5-V supply is shown here, but this is not a rigid requirement. This stage (and most of the others) can operate from supplies of up to 15 V with 3086 or 3046 devices (or higher with the 3146).

The input threshold of this stage is approximately 1.5 V, which means it can be driven from TTL or DTL sources. A LOW input is less than 1.5 V; a HIGH input is greater than 1.5 V.

The HIGH input logic level should be less than 8 V or so to avoid Zener breakdown of Q3 or Q4. Note that any desired number of inputs can be added by using additional diodes (such as C) from a 3039 diode array. If a very high input logic level must be accommodated, 3141 diodes can be used.

This circuit can be modified for the AND function by adding a second inverter following Q5.

A NOR logic gate is shown in Fig. 7-2. This circuit uses the npn transistors as emitter followers to provide the OR function for positive input logic levels. A HIGH input appears at R1 and drives the base of

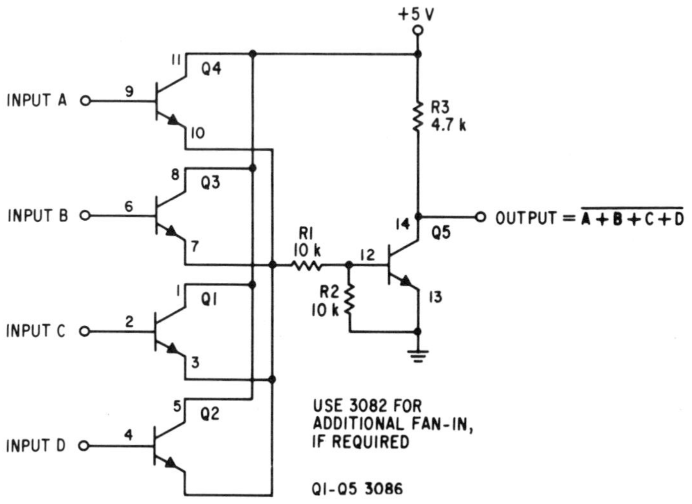

Fig. 7-2 NOR gate

Q5 through the R1–R2 divider. Q5 restores the logic level and inverts the signal, providing an overall NOR function.

The input threshold is approximately 2.0 V, a voltage compatible with most logic types. If desired, this threshold can be lowered by lowering R1 or raising R2.

Supply voltage of this stage is also not critical, but a point to be noted is that it must be *greater* than the HIGH level of the source; otherwise, Q4 (or the other inputs) will conduct collector-to-base. Unused inputs of a gate such as this should be terminated LOW (grounded).

A modification that converts this stage into the OR function is shown in Fig. 7-3; it is realized by adding a second inverting stage.

A logical extension of the two gate types discussed thus far is to combine them in a two-level gating scheme. The circuit of Fig. 7-4 that does this is an AND-OR-INVERT structure. This circuit logically OR's the AND'ed signals A-B and C-D and then inverts the result.

Although not directly realizable with a single IC package, the circuit is shown as an example of what can be done with the proper combination of techniques discussed above.

The above circuits are best suited to TTL or DTL drive sources because of their relatively low input thresholds. However, if desired, they can be readily modified to operate with higher thresholds. Relatively high thresholds are typical of both CMOS (at higher supply voltages) and industrial-type high-threshold logic.

Such a stage is shown in Fig. 7-5. In this circuit, the input threshold is raised to 7 V by connecting a Zener diode (Q4) in the base drive path to inverter Q5. Q5 can turn on only if the input is higher than 7 V, when the current in R1 is diverted to Q5. For lower input levels, Q4 and Q5 are off.

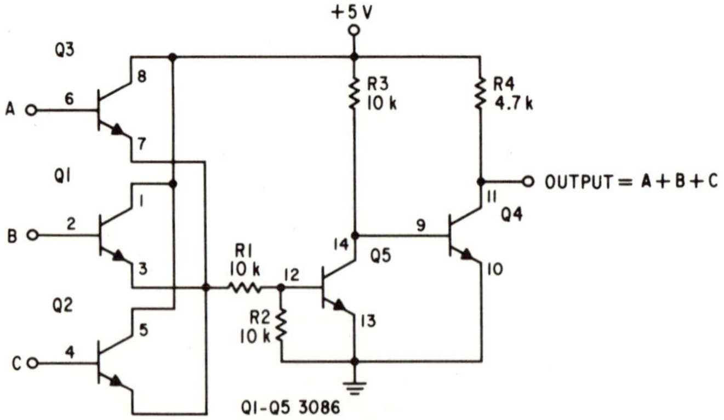

Fig. 7-3 OR gate

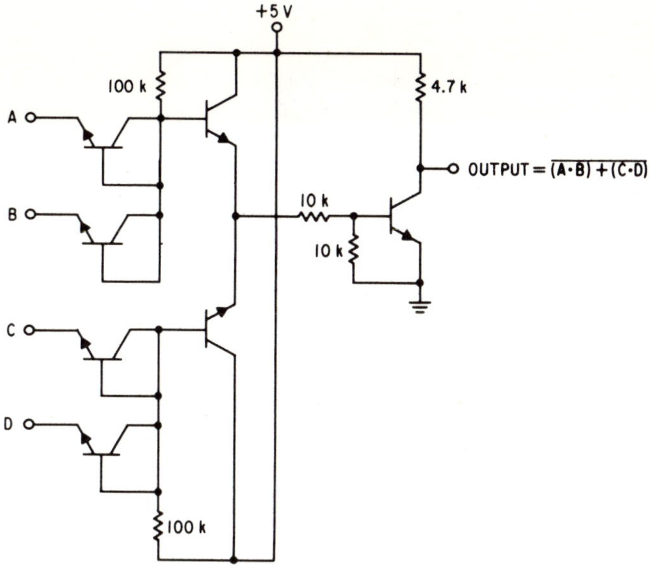

Fig. 7-4 *AND-OR-INVERT gate*

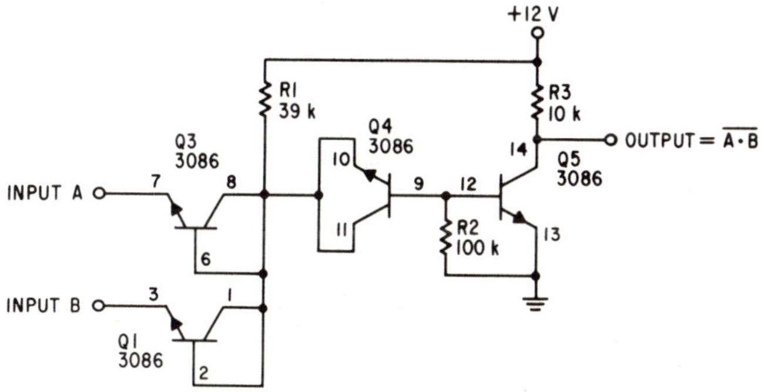

Fig. 7-5 *High-threshold logic stage (NAND function)*

This stage can readily interface with high-threshold bipolar logic or CMOS sources operating from 10- to 15-V supplies. The supply voltage used should be in the 10- to 15-V range. The circuit shown is a NAND function, but other logic implementations are possible as well.

Buffers

The circuits described thus far have been capable of moderate output currents—on the order of a few milliamps. The circuits of Fig. 7-6 are useful as general-purpose buffers or drivers; they can sink 10 mA or more of output current.

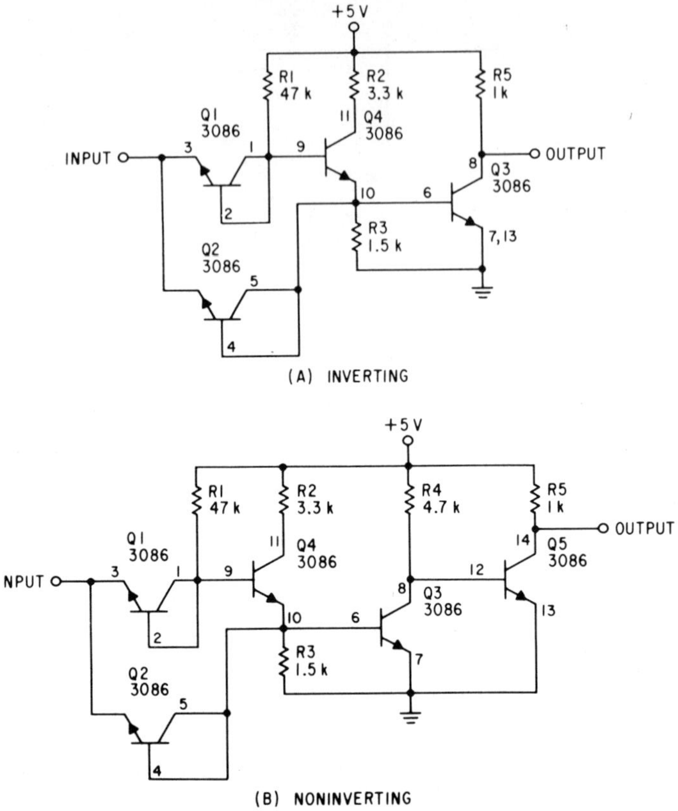

Fig. 7-6 *Buffer/driver currents: (a) inverting stage, and (b) noninverting stage*

Figure 7-6(a) shows an inverting stage; Fig. 7-6(b), a noninverting stage. Both circuits can sink 10 mA of current with a typical 0.25-V saturation voltage; their input threshold is 0.8 V. Appreciably more output-current sink capability can be obtained, if desired, by repinning the circuit to accept a 3083 array.

One-Shot Multivibrator.

A circuit that performs the function of a one-shot (or monostable) multivibrator is shown in Fig. 7-7.

In this circuit's stable or standby state, Q3 is saturated on, and Q2 is off. Consequently, the collector of Q3 is LOW, and Q2 is HIGH. A positive-going trigger pulse to C1–R1 saturates Q1 and pulls the collector of Q1 and Q2 LOW. This negative step is coupled through C_t and

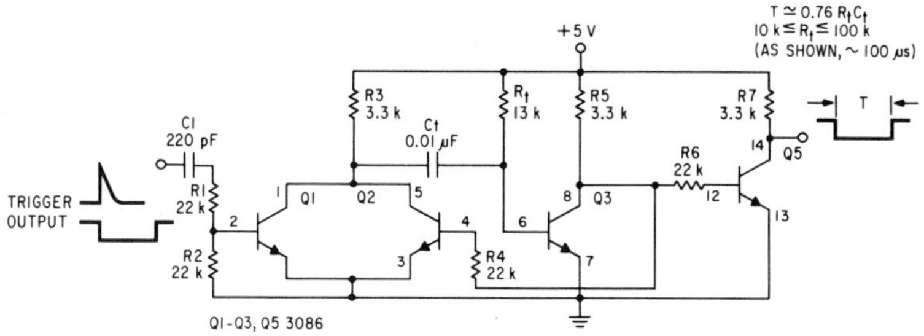

Fig. 7-7 One-shot multivibrator

cuts off Q3, causing its collector to rise. The rise in voltage at Q3 is coupled to the base of Q2 through R4 and regeneratively locks Q2 on after the trigger pulse has ended.

The negative step at the base of Q3 has a peak voltage of −5 V, and thereafter C_t begins charging towards +5 V. The time it takes to reach +0.6 V (the base potential of Q3) is the time duration of the output pulse (T), equal to 0.76 R_t C_t. With the collector of Q3 high during this period, the collector of Q5 is low and acts as the output pulse of the circuit.

The pulse width T is 100 μs for the R_tC_t values shown. For other pulse widths, R_t can be varied over the range shown, whereas C_t can be virtually any value above about 50 pF (the minimum value for minimizing stray effects). Output pulse width should be greater than 10 μs for best predictability, but there is no upper limit to pulse width (other than practical values for C_t). The trigger pulse width must be less than width T for the circuit to function properly; the input differentiator formed by C1–R1 guarantees this requirement.

R-S Flip-Flop

An R-S flip-flop (or latch) circuit is shown in Fig. 7-8. This circuit is simply formed from two cross-coupled inverters, Q2 and Q3. A HIGH state at one collector forces the opposite side to the LOW state. The output from Q2 is defined as Q; it is HIGH in the true state. The output from Q3 is $\overline{Q}$; it is the complement of Q.

The circuit is parallel-triggered by transistors Q1 and Q5, which accept the R and S inputs, respectively. A HIGH (true) input to S sets the Q output to HIGH and the $\overline{Q}$ output to LOW. A subsequent HIGH input to the R input resets the Q output to LOW and the $\overline{Q}$ output to HIGH. The circuit will retain the information of the *last* R (or S) input indefinitely. Although a +5-V supply is indicated, the circuit can accommodate supplies of up to +15 V.

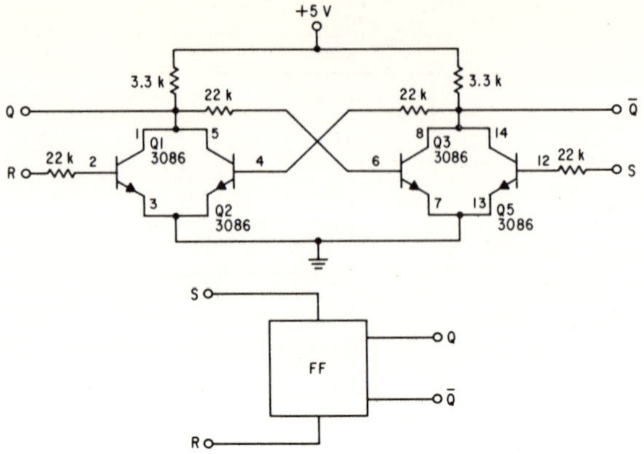

Fig. 7-8 *R-S flip-flop*

Logic Interfaces

It is often required that one form of logic be reliably interfaced with another or that some arbitrary levels be converted to standard logic swings. IC arrays can efficiently handle these interfaces because of their inherent flexibility. Some of the interfaces possible are shown in Fig. 7-9.

Figure 7-9(a) shows a TTL-to-CMOS interface that can drive CMOS stages operating from supplies of +5 to +15 V. To accomplish the TTL input compatibility, a 1.4-V threshold is set up by the Q1–Q4 and Q3

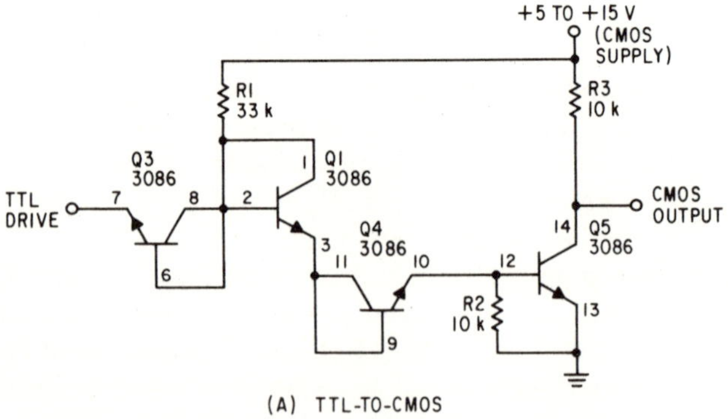

Fig. 7-9 *Logic interfaces: (a) TTL-to-CMOS, (b) bipolar-to-TTL, (c) TTL-to-bipolar, and (d) totem pole driver*

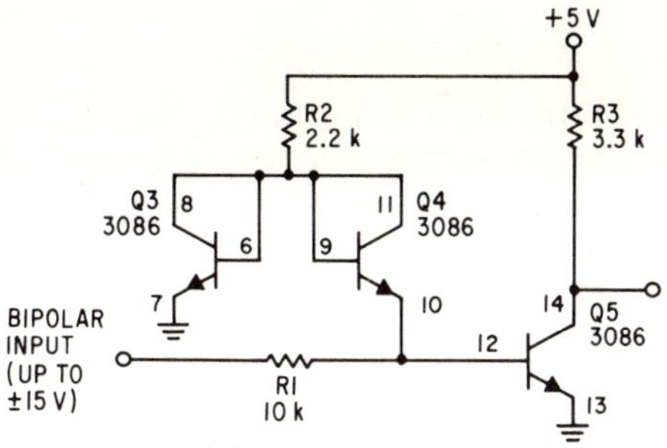

(B) BIPOLAR-TO-TTL

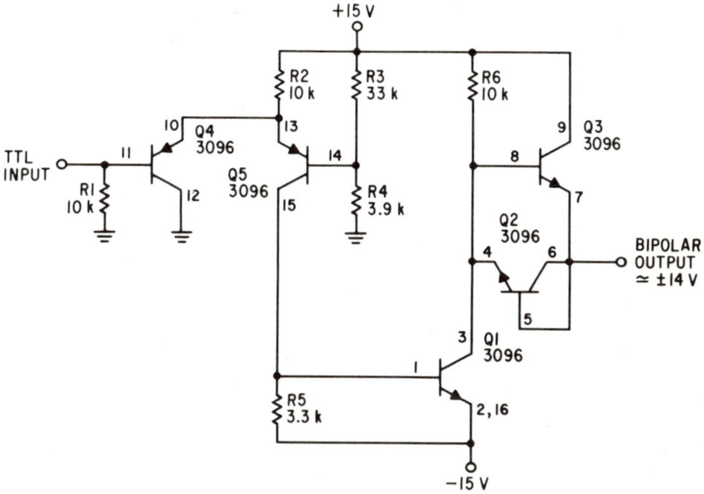

(C) TTL-TO-BIPOLAR

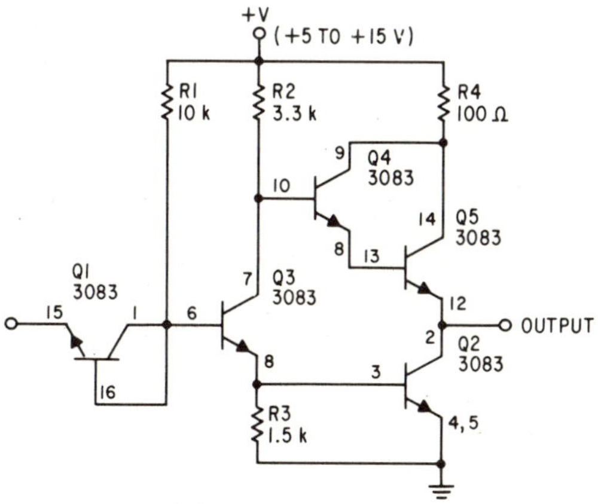

(D) TOTEM-POLE DRIVER

diode connections that drive Q5. This threshold remains essentially constant over the supply range shown.

Figure 7-9(b) illustrates how a bipolar level signal such as is typical of analog outputs can be transposed to TTL levels. A positive input saturates Q5, delivering a LOW output. A negative input cuts off Q5, resulting in a HIGH output. Q3 and Q4 form a diode clamp for Q5 and prevent Zener breakdown. The stage can also be used for CMOS output levels by increasing R2–R3 values.

The circuit of Fig. 7-9(c) provides the reverse form of interface, delivering ±14-V output levels from a TTL input drive. In this circuit, Q4 and Q5 are an input differential pair that provide a current drive to Q1. Q5 is referenced by R3–R4 to a 1.5-V logic threshold that provides the TTL compatibility.

For a LOW logic input, Q4 is on and Q5 is off, as is Q1. The output for this condition will be at the V+ level minus 1 V_{BE}, or about +14 V. For a HIGH logic input, Q4 is off and Q5–Q1 are on. Q1 pulls the output to V− plus 1 V_{BE}, or about − 14 V. Other input logic thresholds can be accommodated as well by adjusting the threshold voltage as desired. R2 should also be adjusted to provide 1.5 mA of current to Q5. When R4 = 33 k and R2 = 4.7 k, for instance, the circuit is well suited to 15-V CMOS drive levels. Note that because of the use of lateral PNP's Q4–Q5, the input can safely accommodate swings in excess of V+. All the transistors are available in a single array, the 3096. The output current capability is ±25 mA.

A high-current, "totem pole" driver stage is shown in Fig. 7-9(d). This circuit can sink 50 mA or more at typical saturation voltages of 0.4 V. A Darlington connection is used for the active pullup Q4–Q5, providing a high-current drive for source currents as well. This gives the circuit excellent capacitive load drive capability.

The input threshold of the circuit, 0.8 V, is relatively constant with supply voltage, a feature that makes the circuit useful as a low-voltage to high-voltage level shifter as well.

Power-Up Reset/Delay

A common requirement in many systems is the "initialize" delay, which provides a single reset pulse, when power is applied, to force logic circuitry into a desired stage. A simple array circuit to accomplish this objective is shown in Fig. 7-10.

In this circuit, Q4 and Q5 of a 3086 (or 3046) are a buffer and inverter, respectively. Q5 is saturated low for Q4 input voltages of more than 3 V_{BE} (~2V). When power is first applied, there is no charge on C1, and Q4 sees an instantaneous input of +5 V. As C1 charges, the voltage across R1 reduces, and Q5 drops out of conduction when the voltage

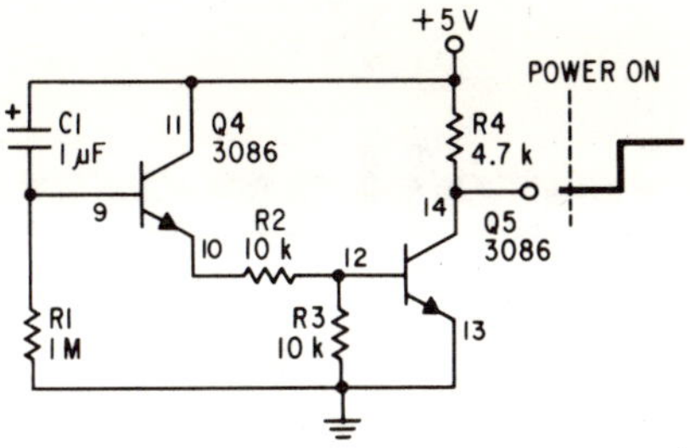

Fig. 7-10 *Power-up reset/delay*

across R1 drops below 2 V. As can be noted from the timing sketch, this condition holds the collector of Q5 low for the reset delay timing period, after which it rises. For the values shown, the delay is 0.5 s, and timing can be adjusted via either R1 or C1.

REFERENCES FOR CHAPTER 7

(1) ———, "The TTL Data Book," 2nd ed., Texas Instruments, Dallas, Texas.

(2) Havasy, A., and Kutzin, M., "Interfacing COS/MOS With Other Logic Families," RCA ICAN-6602, Nov. 1973.

8
Miscellaneous Circuits

This final chapter consists of a collection of circuit types that, for one reason or another, do not readily fit into the broad topic areas of the other chapters. Although such a grouping may be viewed as a catch-all category, it nevertheless contains a variety of useful circuits that exploit IC array features, often to very good advantage.

Detectors

Signal detectors and/or rectifiers are a very common circuit requirement of considerable variety. Basically, the concept is to convert some ac signal into a corresponding dc voltage equivalent. Since most readers are probably very familiar with the fundamentals of this process, it will not be repeated here. What will be covered are more refined and flexible versions of detectors and rectifiers that offer increased performance as a result of IC array techniques.

A good circuit example that illustrates how the characteristics of IC arrays can be put to use is the simple AM detector found in virtually all receivers. When an ordinary signal diode is used to AM-detect an RF carrier, the circuit is subject to a basic lower threshold below which the carrier is either not detected or detected nonlinearly. This results in loss of dynamic range or distortion, or both.

Both problems can be simplified or eliminated by using a biased detector such as that shown in Fig. 8-1. Here, a pair of IC array diodes are used, with D6 developing a dc bias that is applied to the anode of D1 through the secondary of T1. Since D1 and D6 are matched, D1 is at the threshold of conduction, even with no RF input. When an RF signal is applied, the linear detection range of D1 is extended to very low levels, and distortion is reduced considerably. Note that in implementing this

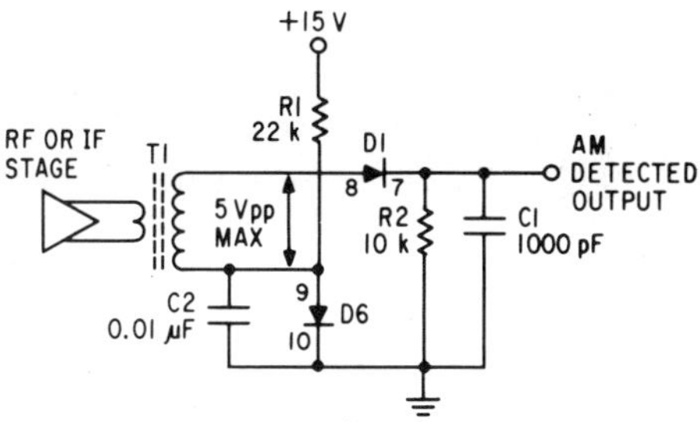

Fig. 8-1 *Biased AM detector*

circuit the breakdown voltage of D1 must be observed; for this purpose, the pp swing across T1 should be 5 V or less.

The biased rectifier concept is not limited to transformer-coupled circuits and can be extended to more general circuit applications as well. An example of a general-purpose biased detector is shown in Fig. 8-2. In this circuit, D1 furnishes dc bias to the detector as before, but the detector in this case is an npn transistor, Q1. Using a transistor instead of a diode has the useful advantage of the former's forward gain and therefore loads the source less. The signal is ac-coupled by C1, and the dc bias to Q1 is applied by resistor R1.

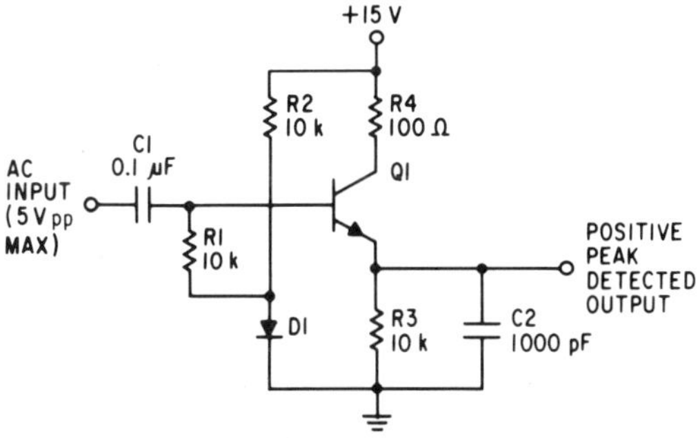

Fig. 8-2 *General-purpose biased detector*

The selection of R1 should be such as to minimize loading on the source without introducing excessive dc drop in the base path of Q1. As a necessity then, this resistor is somewhat of a compromise in terms of bias when compared to the transformer's low dc resistance. Typical values might be in the range of 10 to 100k.

On positive signal peaks, C2 is charged by Q1, and the detector's decay time constant is formed by R3–C2. This circuit detects positive peaks as previously shown; it is also limited to 5 V pp max signals.

A limitation of the peak-detecting circuits described thus far is that they are limited in the dynamic range of the input signal they can handle by the reverse base-emitter voltage of the detector transistor (or diode). The circuit of Fig. 8-3 gets around this limitation rather neatly. It can handle signals up to the V_{CB} rating of the transistors, which in this case are 3096 types. This circuit, which is called a wide-range peak detector, also has a biased detection threshold but one that is set up in a different manner.

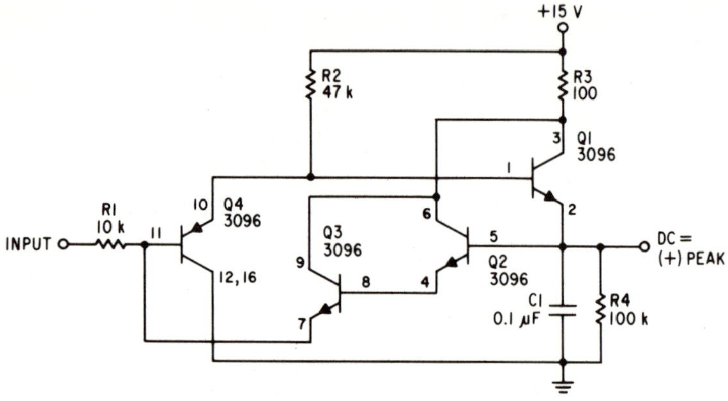

Fig. 8-3 Wide-range peak detector

Here, pnp Q4 drives npn Q1, which serves as the detector for positive peaks, storing a change on capacitor C1. Since Q4 and Q1 form a complementary emitter follower, their V_{BE}'s essentially cancel. Thus the positive peak level seen at the input is stored on C1.

When the input passes the positive peak and goes negative, Q1 sees a reverse V_{BE} as a result of the charge on C1. If the voltage at the base of Q1 or Q4 is not limited in peak negative swing, Q1 will break down emitter to base and most likely fail. In the circuit here, however, Darlington npn pair Q2–Q3 buffers the voltage on C1 and provides a low-impedance negative clamping potential to the base of Q4. With this provision, the reverse voltage of Q1 is never more than 1 V_{BE}.

The frequency range of this circuit is not as wide as that of the others because of the use of the pnp. However, it is useful up to about 100 kHz and at voltage levels of 15 V or more.

Full-Wave Rectifiers

The AM detectors and peak detectors just described are half-wave forms of signal rectifiers. Often, requirements dictate the need for a full-wave rectifier in which both alternations of signal input are converted to dc. The matched characteristics of IC arrays come into useful play in these circuits as well.

A precision full-wave ac-to-dc converter circuit is shown in Fig. 8-4. This circuit accepts ac inputs of up to ±10 V (7 V rms) over a 10-Hz to 100-kHz range and converts them to a proportionally scaled, positive dc voltage. It consists of two elements: a rectifier and a current-to-voltage converter. A1 is a dual op amp, the sections of which are shared by the two elements.

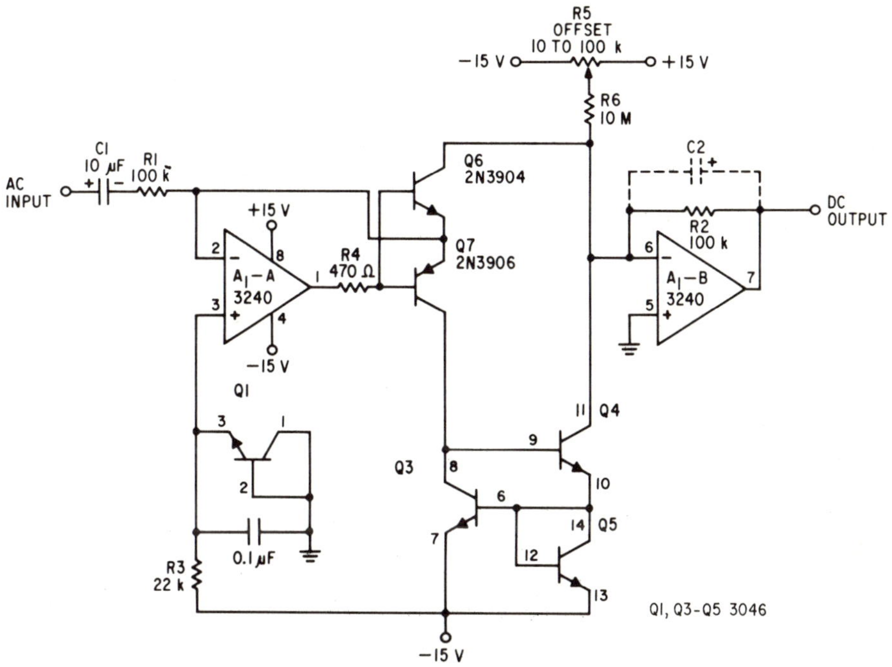

Fig. 8-4 Full-wave ac-dc converter

A1–A and Q1–Q2 form the rectifier part of the overall circuit.[1] The actual rectification takes place in Q6 and Q7, a pair of complementary, high-gain discrete transistors. For negative-going inputs to C1, Q6 develops a negative-going current in its collector. For positive-going inputs, Q7 develops a positive-going current into the Wilson current mirror formed by Q3–Q5 of a 3046. The output from Q4 is a negative-going current for positive-going inputs.

The net result of these conditions is that both positive and negative signal alternations develop the *same* polarity of current output. This is

[1] See C. C. Todd and H. Stellrecht, Ref. (9).

by definition *full-wave rectification*. The rectified current is fed to A1-B, a current-to-voltage converter, where it is reconverted to a full-wave rectified voltage.

In the rectifier portion of the circuit, there are several items that are keys to the operation. The precision of the rectification is to a large extent dependent upon the op amp used to drive Q6–Q7, and the type employed should preferably have a high slew rate and bandwidth for good accuracy to high frequencies. Also, it should have a low bias current so as to not limit the dynamic range at low signal levels. The type specified is a FET input; others also useful are the TL072 and LF353. With somewhat decreased performance, a dual 741 type such as the 1458 or 4558 can be used as well.

Overall sensitivity of the circuit can be scaled by feedback resistor R2 for different full-scale levels. As shown, the output will be +10 V dc for a (+) or (−) 10-V-ac input peak, or unity gain. The output can be filtered easily, if desired, by shunting R2 with an averaging capacitor, C2. The value of this capacitor can be optimized for minimum ripple. If the averaged dc output level is going to be scaled to be equal to the RMS value of the input, then the value of R2 should be multiplied by 1.1 (or 110k).

Either R1 or R2 can be trimmed for scale-factor calibration, since both affect the conversion gain. Accuracy at levels of 100 mV or less will be limited by the offset of the output op amp. This can be trimmed out, if objectionable, by using the R5–R6 network. R5 should be adjusted for a zero dc output with no ac input to C1.

Precision AC Metering Circuit

Figure 8-5 shows a circuit that can be used to convert an ac voltage to a proportionally scaled *current* that drives a dc meter. It is suitable to a

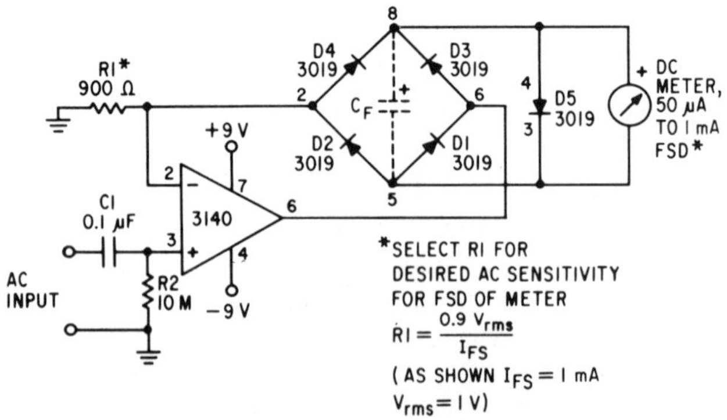

Fig. 8-5 Precision ac metering current

wide range of meter sensitivities and can be easily scaled to match various ac signal voltage inputs for full-scale deflection.

In this circuit, the op amp is used as a precision voltage-to-current converter, converting the ac input voltage appearing across R2 to a proportional current in R1. Since the meter is within a diode bridge and within the feedback path, the bridge converts the ac current to dc in the floating meter. Note that because the bridge and meter are within the overall feedback loop, the temperature sensitivity and nonlinearity of the diodes are of small consequence and do not affect stability or calibration.

The circuit is scaled to match a given meter movement and input level by an appropriate choice of R1. For example, with 1 V rms in and a 1-mA FS meter, the equation indicates that R1 should be 900 Ω. Note that the factor of 0.9 is used to scale between rms ac and the average dc value, which the meter reads.

A 3019 diode bridge serves nicely as the rectifier; as a bonus, it has a shunt diode, D5, that can be used for overload protection. Some shunt filter capacitance may be necessary at very low frequencies where the mechanical time constant of the meter may no longer average the dc properly; C_F can be chosen for such an occasion.

This circuit can be battery-operated, if desired, as the entire current drain is typically 4 mA. Higher input voltages can be accommodated by appropriate attenuation prior to C1.

Gain-Control Circuits

A very wide variety of circuits that can be directly gain controlled by dc (or ac) signals are possible using bipolar transistors. Many of these circuits are based in some manner on the principles of multiplication described by Gilbert,[2] which are themselves based on the fundamental amplification properties of the transistor. Efficient, high-performance multipliers based on the bipolar transistor require matched transistors, which, of course, are a natural product of the IC process. Thus, IC array devices can very effectively be used to implement the concepts of multipliers and gain-control circuitry.

A simple multiplier circuit that is useful for demonstrating the gain-controlled principle is shown in Fig. 8-6. In this circuit, Q1–Q2 is the gain-controlled pair and operates differentially. Q1–Q2 is fed an emitter current, I, from the controlled source, which is made up of Q4–Q5 and A2 of the dual op amp.

In Chap. 2 is was shown that the gain of a differential pair such as Q1–Q2, when differentially loaded, is

$$A_{VD} = 19.3\ I_E R_L$$

[2] See B. Gilbert, Ref. (3).

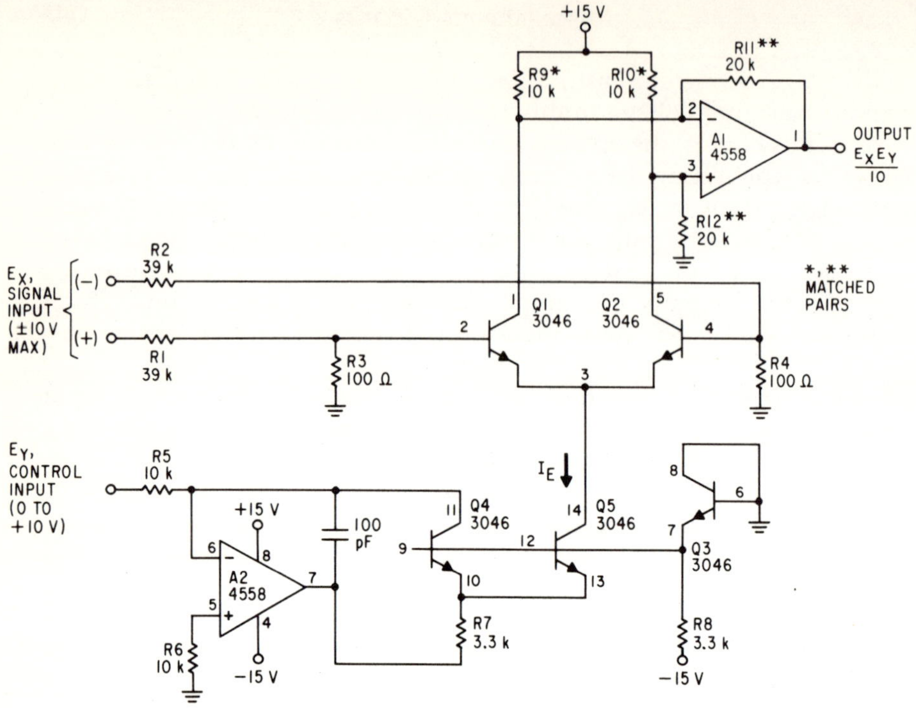

Fig. 8-6 Basic two-quadrant multiplier gain control

In the case here, R_L = R11 = R12. Since I_E is variable as a function of the control input, the gain of Q1–Q2 is proportional to the control voltage, E_y. The output is then the product of the input E_x and control E_y, divided by a constant. It is mathematically expressed as shown in the figure.

Two other factors remain to make the circuit practical. One is that the control signal cannot be allowed to appear in the output or it will cause transients as the gain is changed. It is rejected (here) by the matched resistor bridge made up of R9–R10 and R11–R12 and by the second half of the op amp. The second problem is due to the very high gain of the differential pair; it is linear only to signal levels up to about 20 mV. The input attenuators R1–R3 and R2–R4 serve to reduce a higher input level down to a range compatible with the operating region of the Q1–Q2 linear range.

For the values shown, I_E is a maximum of 1 mA when E_y is +10 V, and this represents the maximum circuit gain condition. The gain from Q1–Q2 to the output is 400 for this condition; thus, the input attenuators are chosen for this attenuation. A 20-V-pp signal can be handled at the E_x input; it may be applied in either a single-ended or differential form, with polarity as noted. The gain-control range is 80 dB or more, and the widest control range will result when the A2 op amp has a low offset voltage and low bias current. The amplifier type shown is a general-purpose one; higher performance can be realized with some specialized units (described in the following circuit).

One of the weak performance points of this circuit is its distortion, which is on the order of 2 percent at the full output level of ±10 V vms. This is due to the nonlinear transconductance of the differential pair with increasing signal input levels [see Fig. 2-20(b)]. The balanced nature of the differential operation supresses *even* order distortion, but odd order distortion (mostly third harmonic) is still significant at high input levels.

The circuit is still useful for many applications, however, and is attractive for its simplicity. Bandwidth is limited in the form shown by the output op amp. Other forms of differential-to-single-ended conversion can be used at the Q1–Q2 collectors if desired, such as a center-tapped transformer. The basic amplifier gain-control mechanism that exists in Q1–Q2 is useful from dc to VHF frequencies and in most circuits will be limited primarily by the peripheral circuits used.

An improved form of gain controller is shown in Fig. 8-7. This circuit, which evolves directly from the previous one, features a substantial performance increase in linearity because of a system of nonlinearity compensation.

In Fig. 8-6, it was mentioned that the relatively high inherent gain of the multiplier circuit must be preceded by a complementary attenuator. If this attenuator is a simple resistive one, the nonlinearity of the differential pair will be serious at high signal input levels.

In this circuit, the input attenuator is composed of R1 in series and a 3019 diode bridge shunt. These diodes, biased at a 500-μA level by R2–R3, constitute a 100-Ω dynamic impedance. Their actual instantaneous impedance follows the input current, however, in such a manner as to produce a logarithmic signal voltage. This voltage is the complement of that required by Q1–Q2 for linearization, and the net result is a much more linear transfer characteristic.

The use of the linearization diodes allows a greater input signal to be applied before distortion becomes serious; thus, the maximum control current of this circuit is reduced, to 625 μA at +10 V, with R5 equal to 16k. With the increased linearity available, an offset trim is desirable for minimum distortion. This can be accomplished by using the (optional) network R14 and R15. Distortion of this circuit is less than 0.1 percent after trim, at output levels up to 5 V rms or more. By contrast, the simple circuit without linearization has a distortion of about 5 percent for similar conditions.

The circuit can also be used as an AM modulator, if desired, by using a simple bias option. If a bias resistor R4 is added to the input of A2 to operate at one-half gain, a ±10-V-ac signal at E_y can then be used as a modulation input (with E_x constant, as a carrier). For the circuit values shown, the correct value of R4 is 47k.

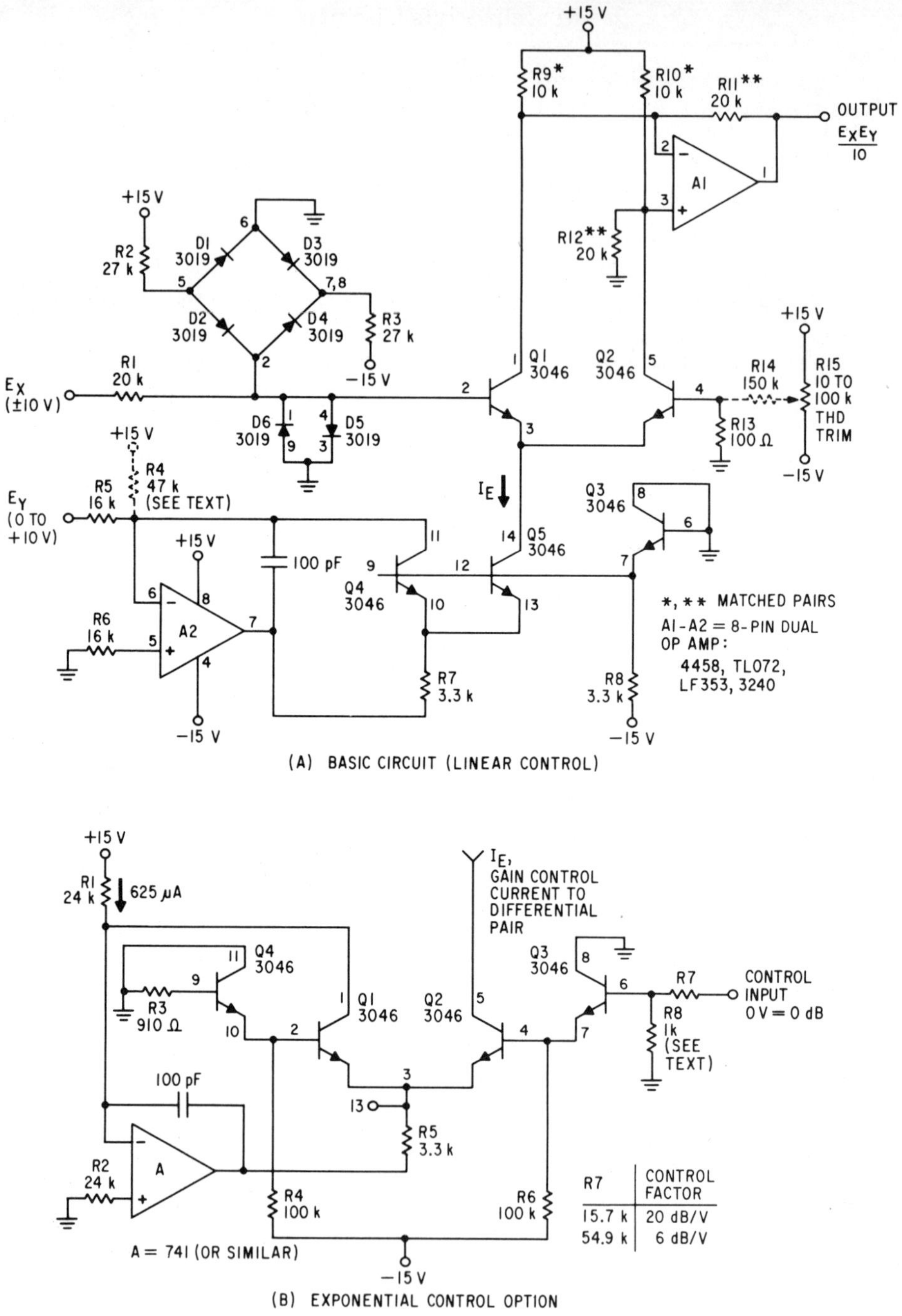

Fig. 8-7 *Linearized gain control: (a) basic circuit, and (b) exponential control option*

A number of dual op amp types are listed in the notes. Of those noted, widest bandwidth will be realized with the FET input devices: TL072, LF353, and CA3240.

Exponential-Control Law Circuit

With only slight additional complexity, the gain-control circuit of Fig. 8-7(a) can be adapted to a form of *exponential* control as opposed to linear control. Exponential control allows a linear variation in control voltage to change the output a specified number of decibels. A modification that can be added to the circuit of Fig. 8-7(a) is shown in Fig. 8-7(b).

This circuit replaces A2 and Q4–Q6 of Fig. 8-7(a) and develops a current I that is fed to Q1–Q2 in the previous circuit. The current I is exponentially related to the control voltage and has a scale factor (or sensitivity) of 20 dB of gain control per volt of control. Zero voltage represents maximum gain (which is unity), and −5 V, the minimum gain of −100 dB.

The circuit is an antilog generator with a buffered input, which drives a matched pair from a 3046 array, Q1–Q2.[3] Q1 is operated at the full-scale reference current of 625 μA by virtue of R1's connection to +15 V. With zero control voltage, Q1 and Q2 see the same V_{BE}; thus, Q2 develops a 625-μA output current for full gain. As the control input goes negative by 1 V, the R7–R8 divider produces a 60-mV change at Q3 and also at the base of Q2. For each 60 mV of change in V_{BE}, the collector current of Q2 changes 10/1 (see Chaps. 1 and 2), which provides 20 dB of attenuation.

R7 can be selected to scale the 60 mV/decade seen at Q3 to any desired sensitivity, as the application may dictate. For example, it can be a value of 54.9 k (see table) if a sensitivity of 6 db/V is desired. R8 should (ideally) have a TC of 0.3 percent per °C to compensate for the gain variation of the transistors if stability in wide temperature ranges is desired. Alternatively, if high stability is desired, Q1–Q2 can be a temperature-stabilized pair (see Chaps. 4 and 5).

Switched Amplifiers

In addition to the gain-controlled amplifiers just discussed, another variety of specialized amplifier is the switched (or gated) amplifier. Useful for on-off control, switched amplifiers using IC array devices provide a number of desirable characteristics, among them fast switching, very high on-off isolation, good linearity, and the ability to provide buffering at input and output as well as gain.

---

[3] See D. Sheingold, Ref. (4), and W. G. Jung, Ref. (5).

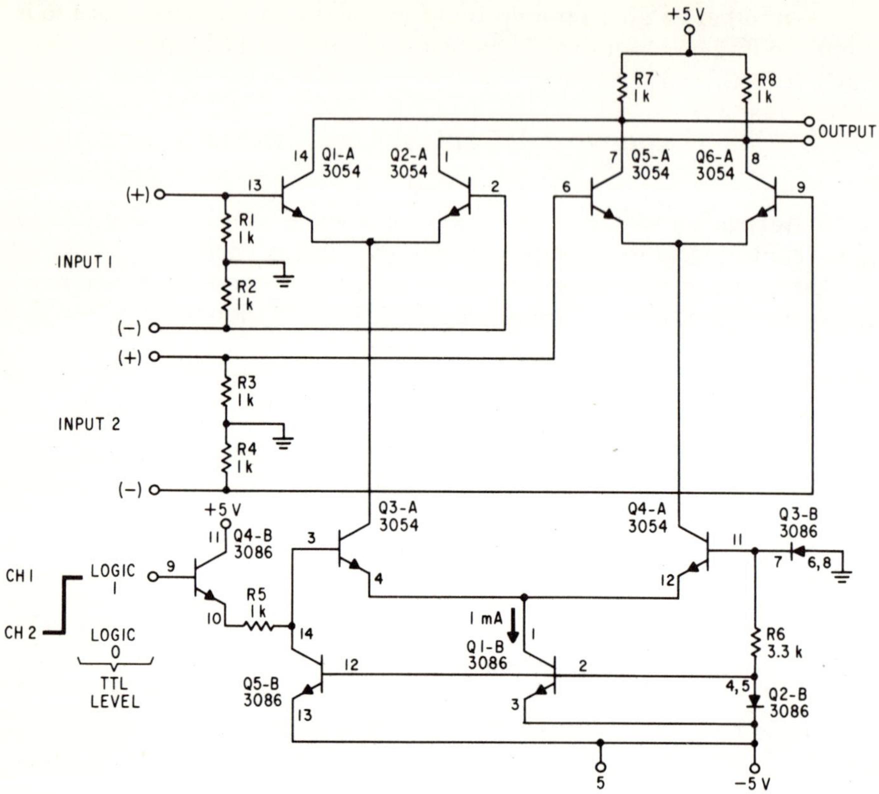

Fig. 8-8 Dual-differential switched amplifier

The circuits described in this section are valuable not only in themselves but also in combination with other array circuits for even broader applications.

Dual-Differential Switched Amplifier

Figure 8-8 illustrates a wideband amplifier that can be used to select one of two differential input signals.[4] Inputs 1 and 2 are applied to two differential pairs, Q1–Q2 (A) and Q5–Q6 (A). At any given time, only one of these two pairs are on, that is, Q1–Q2 (A) or Q5–Q6 (A).

The channel that is to be active is selected by a third differential pair, Q3 –Q4 (A). This pair is driven by a logic-control input via Q4(b). When the control input is HIGH, Q3 (A) and Q1 –Q2 (A) are on and input 1 is active. When the control input is LOW, Q4 (A) and Q5–Q6 (A) are on and input 2 is active.

[4] See Ref. (21).

The switch is composed of a 3054 array (A), whereas the drive to it is a 3086 array (B). Bias to the amplifier is controlled by R6 and the −5-V supply; as shown, it is set at 1 mA. This gives a voltage gain of 19.3 for the active channel.

The circuit as shown is capable of wide bandwidth, since the load resistance is low (1 k) as long as load capacitance is minimized. Higher gain can be realized by increasing the load resistors and positive supply voltage (observing ratings, of course). Although both differential inputs and outputs are shown, the circuit need not necessarily be completely differential. A single-ended output (at half the gain) is available between each output terminal and ground, for example. Choice of the appropriate terminal allows polarity selection.

Single-Ended Switched Amplifier[5]

The circuit of Fig. 8-9 operates in a slightly different fashion, performing the switching function after amplification. This circuit selects between two single-ended inputs and steers each of them to one of two outputs with complementary switch states.

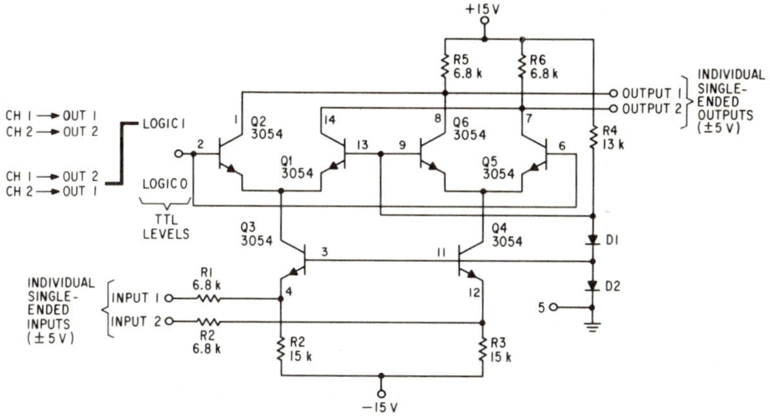

Fig. 8-9 *Single-ended switched amplifier*

Here, the switches are Q1–Q2 and Q5–Q6 of a 3054, with Q3 and Q4 of the same device performing as common-base amplifiers. Q3–Q4 are biased at their bases at 1 V, allowing input signals to be dc-coupled with low offset. The outputs of Q1–Q2 and Q3–Q6 are coupled in such a manner that load resistors R5 and R6 receive either (only) the current of Q3 or that of Q4, or vice-versa.

[5] For a more detailed description, see Ref. (11).

A logic control signal to the bases of Q2–Q5, with Q1–Q6 biased to 1.2 V, allows a TTL compatible drive, with a control sense as noted.

Both inputs and outputs of this circuit are single-ended and can handle levels up to ±5 V. Because the circuit operates entirely as a common base cascode, there is no miller capacitance. This fact plus the wide transistor bandwidths allow operation into the VHF region (if the output time constant is minimized).

The circuit can be used to realize voltage gain, if desired, by reducing the values of R1 (R2) below those of R5 (R6). In general, the voltage gain of a particular channel is R5 (or R6) to R1 (or R2).

Very high isolation is a virtue of this switch, and, if care is taken in the layout, it can be as high 90 dB or more at audio frequencies. Switching time is fast, on the order of 50 ns or less.

Since the circuit has current summing properties at both inputs and outputs, it lends itself to a variety of flexible configurations—bussed outputs or inputs, gain-scaling amplifiers, and so forth.[6]

High-Isolation Series-Shunt Gate

Certain types of switching arrangements offer very high performance with regard to specific performance parameters. For example, if very high off-state isolation is desired, a series-shunt diode gate is very useful. An example of this type of signal switch is shown in Fig. 8-10.

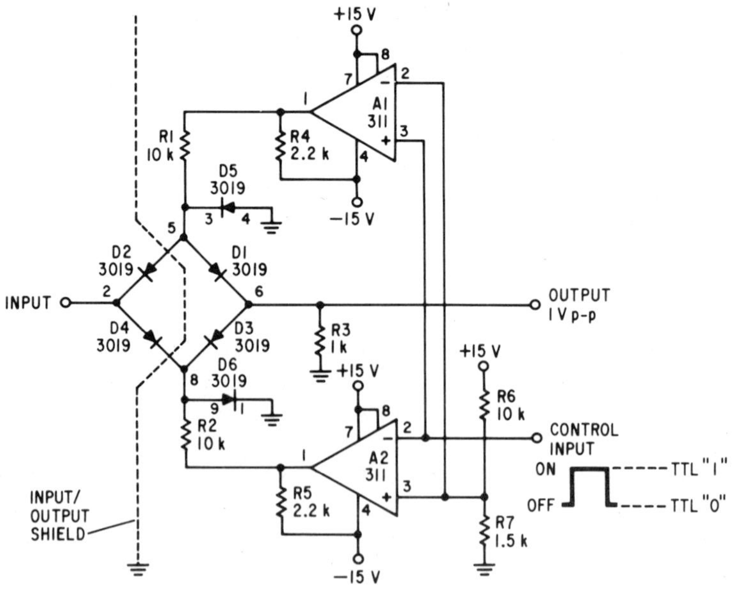

Fig. 8-10 High-isolation series-shunt gate

[6] Many of these are described in Refs. (11) and (12).

In this circuit, a 3019 diode array is used in a switched bridge configuration. Current to the bridge is supplied via resistors R1 and R2, and the output appears across R3. R1 and R2 are driven, in turn, by comparators A1 and A2.

As can be noted from the circuit, the inputs of the two comparators are cross-connected, thus causing their output to be complementary. A TTL HIGH causes A1 to be HIGH and A2, LOW, thereby forcing current in the bridge and back-biasing D5 and D6. The signal is transmitted to R3 with a small insertion loss—about 0.3 dB. Maximum pp output is set by the current in R1 and R2; it is 1 V or less.

When the input logic reverses, the bridge diodes turn off and both D5 and D6 are on. In this off state, bridge diodes and D5—D6 form a T pad with very high isolation—90 dB or more at audio frequencies. This will be enhanced if an input-output ground-plane shield is used to prevent capacitive coupling and high cross talk. For the circuit shown, the switching time is 1 μs or less and is limited by the response time of the comparator. Further description of circuits of this type are contained in Ref. (13).

Limiters, Clippers, and Clamps

It is often necessary in signal and waveform processing to be able to limit, clip, or clamp a waveform predictably to a definite reference level. IC array transistors, because of their matched characteristics, lend themselves quite readily to a variety of such uses.

Zener Limiters

We have already seen how transistor EB junctions can serve well as stable voltage references, but it may not be so obvious that they also serve well as fast, low-TC-signal limiters also. Some examples are shown in Fig. 8-11.

Simply placing a single Zener-connected npn transistor within a diode bridge gives a symmetrical and stable voltage breakdown of 8.4 V, as shown in Fig. 8-11(a) The transistor can be any (single) array EB junction, connected as shown. The diodes cannot be monolithic types, however, because of their low breakdown voltage. A discrete diode bridge should be used; it can be any diode with a 10-V rating or better. Note that this arrangement is low TC, since two forward diodes compensate a single Zener. This arrangement is useful to much higher frequencies than those of the often-used parallel IN750 series types because of lower diode capacitance.

A 3019 diode bridge can be used to obtain a symmetrical breakdown as shown in Fig. 8-11(b). If the current paths from pins 2 to 6 are examined, parallel Zener paths will be seen, as shown in the equivalent

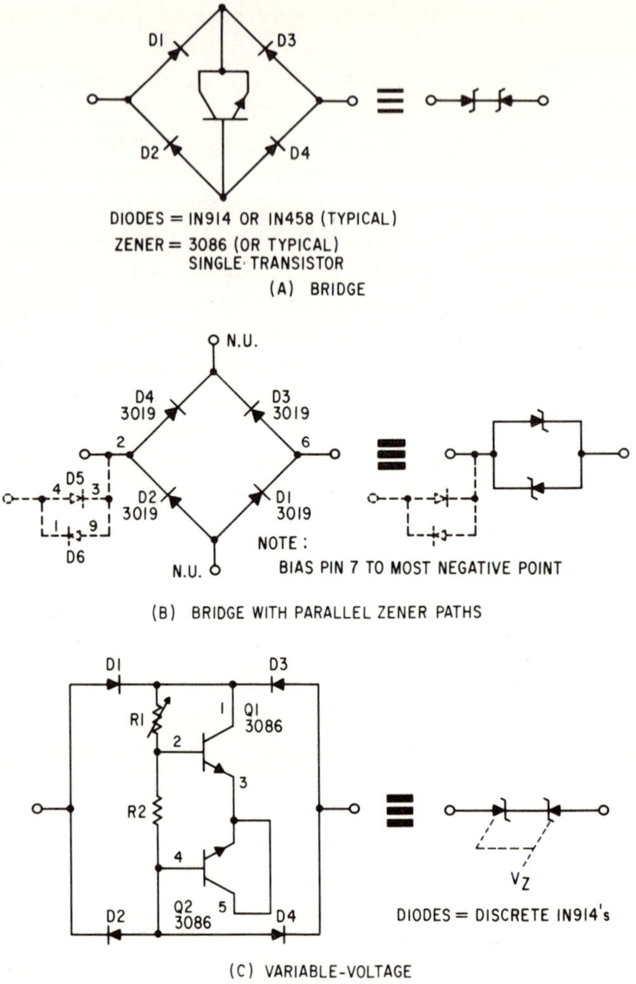

Fig. 8-11 *Bipolar Zener limiters: (a) bridge, (b) bridge with parallel Zener paths, and (c) variable-voltage shunt regulator*

circuit. Since the Zener voltages are matched within a few percent, the symmetry of breakdown will be good at approximately ±7.7 V.

Note that this voltage breakdown is *not* minimum TC, since only one forward diode compensates the Zener. A lower TC can be optionally obtained by connecting D5 and D6 in parallel and the two of them in series with the Zeners, as shown by the broken lines. This will give a ±8.4-V breakdown with a nominally low TC, and the voltages will still be matched.

An interesting variation of the single Zener within a diode bridge is shown in Fig. 8-11(c). Here, Q1–Q2 of a 3086 (or similar) array forms a variable-voltage shunt regulator (see Chap. 4) within the diode bridge formed by D1–D4. R1 programs the Zener limit voltage from about 9 V upwards.

This circuit will have a shunt "leakage," because of R1–R2, for voltages between 1.4 V and the breakdown voltage. The effect of this can be minimized by maintaining R1–R2 at relatively high values. Note that because of the current gain of Q1, this circuit can conduct appreciable current.

Shunt Clippers

A common form of signal clipper is the shunt clipper, often consisting of nothing more than a series resistor and a shunt diode to ground. Although this type of clipper can be easily implemented with IC diodes, other interesting variations of it are possible by making use of transistor action.

In Fig. 8-12 are illustrated two single-ended shunt clippers, using transistors as "programmable diodes." R1 is a series input resistor, and R2–R3 programs Q1 as a V_{BE} multiplier. Q1 will turn on at a voltage equal to (R2 + R3)/R2 times its V_{BE}.

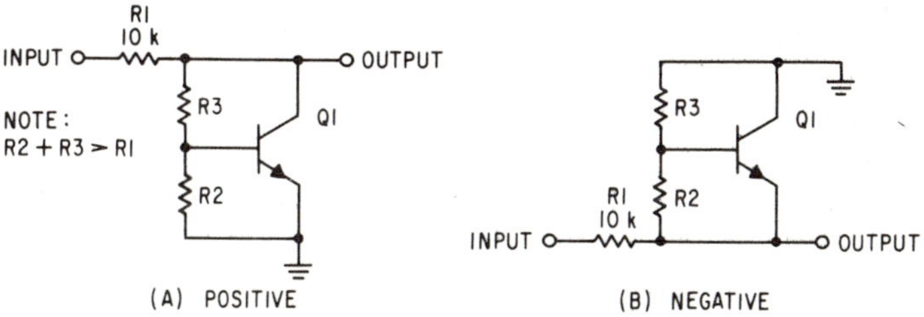

Fig. 8-12 Single-ended shunt clippers: (a) positive, and (b) negative

Either a positive (a) or negative (b) polarity clipper may be implemented, as desired. For best results, the values of R2 and R3 should be higher than that of R1 to minimize the resistive voltage division when Q1 is off.

Note that EB and substrate diode-parasitic action should be avoided with this circuit by controlling the reverse potentials applied to it.

Buffered Variable Clippers

A form of signal clipper that is nearly ideal in its signal properties is the buffered clipper of Fig. 8-13. This circuit offers very abrupt and

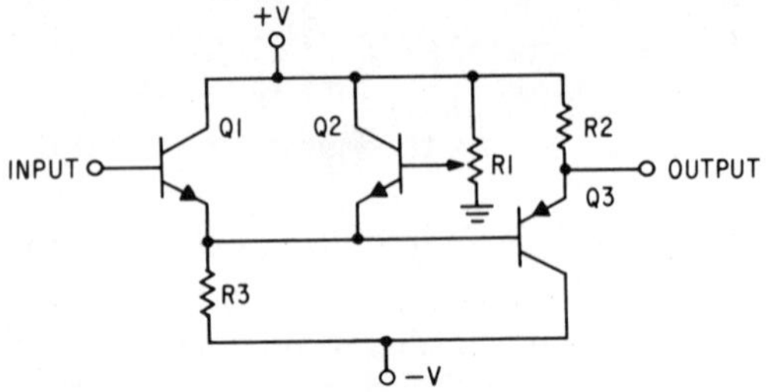

(A) NEGATIVE

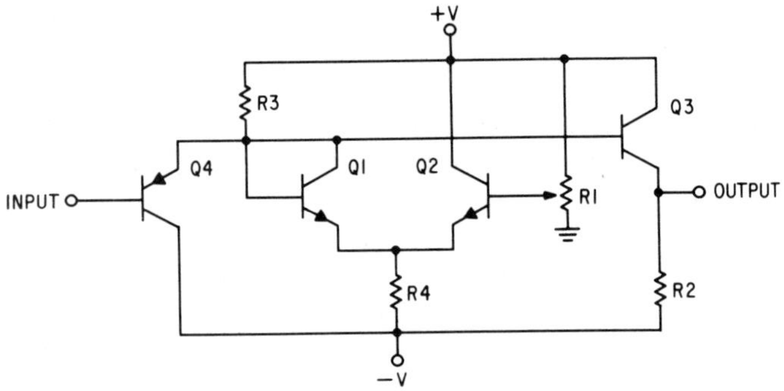

(B) POSITIVE

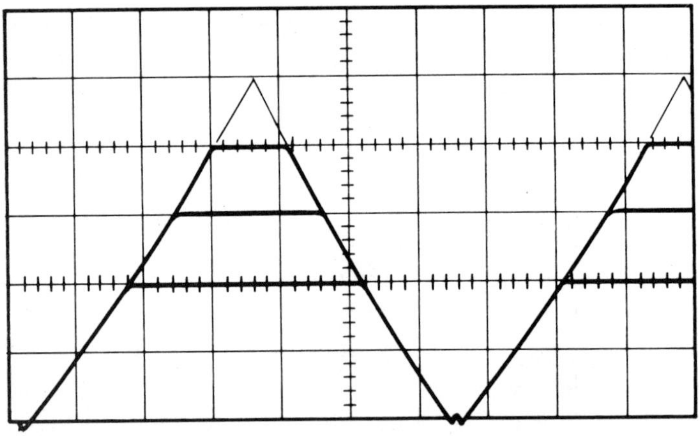

(C) INPUT-OUTPUT WAVEFORMS OF POSITIVE BUFFERED VARIABLE CLIPPER

Fig. 8-13 *Buffered variable clippers: (a) negative, (b) positive, and (c) input-output waveworms of positive clipper, with input a linear triangle wave and output limited at three increasing levels (multiple exposure)*

complete limiting that is temperature-compensated and has buffered inputs and outputs.

In Fig. 8-13(a) is shown a negative form of the clipper, which is basically composed of Q1–Q2. Q1 performs as a single npn emitter follower for signals below the limit threshold and drives pnp follower Q3. The respective V_{BE}'s of these transistors cancel, allowing dc coupling with low offset.

For negative-input signal peaks below the voltage threshold established by limit control R1, Q2 turns on and "robs" the current in R3 away from Q1. This action clips the negative signal peak in very abrupt fashion, since the emitter impedance of Q2 is low and that of Q1 becomes very high for this condition.

Q1 and Q2 are a matched npn pair such as are found in a 3096, which can also include Q3. Alternatively, Q1–Q2 can consist of 3086 or 3046 units, and Q3 can be discrete. Note that this clipper is limited to input levels *less* than the EB breakdown of Q1; practically, this means 5 V pp, or less.

A similarly operating positive limiter is shown in Fig. 8-13(b). Here, Q4–Q3 form a pnp-npn complementary emitter follower for low dc offset. Between these transistors, Q1 and Q2 form a positive clipper that deprives Q4 of emitter current when the voltage at diode-connected Q1 rises above that set by clipper control R1. Again, limiting action is quite abrupt and complete.

The performance of this positive clipper is demonstrated by the multiple-exposure scope photo of Fig. 8-13(c). This is a superposition of an input triangle waveform and three successively increasing output waveforms that have been limited.

Zero-Offset Voltage Follower with Peak Limiting

A high-speed zero offset voltage follower can be constructed by using the balanced properties of an IC array transistor differential pair. By adding current-biased elements, the output voltage swing can be made bidirectional, with a predetermined current limit in both directions of signal swing (current sourcing and current sinking), as in Fig. 8-14.[7]

In this circuit, Q1 and Q2 are a matched differential pair with current-source bias provided by Q4, using 3018 array devices.

Q1–Q2 are optimized for high-frequency response, since Q1 operates as an emitter follower and Q2 as a common base. Furthermore, Q2 and Q3 form a cascode amplifier, with the dynamic collector current of Q2 being transferred to the collector circuit of Q5 by way of the latter's common-base connection. The collector load for Q5 is current source Q.

[7] For source, see W. G. Jung, Ref. (15).

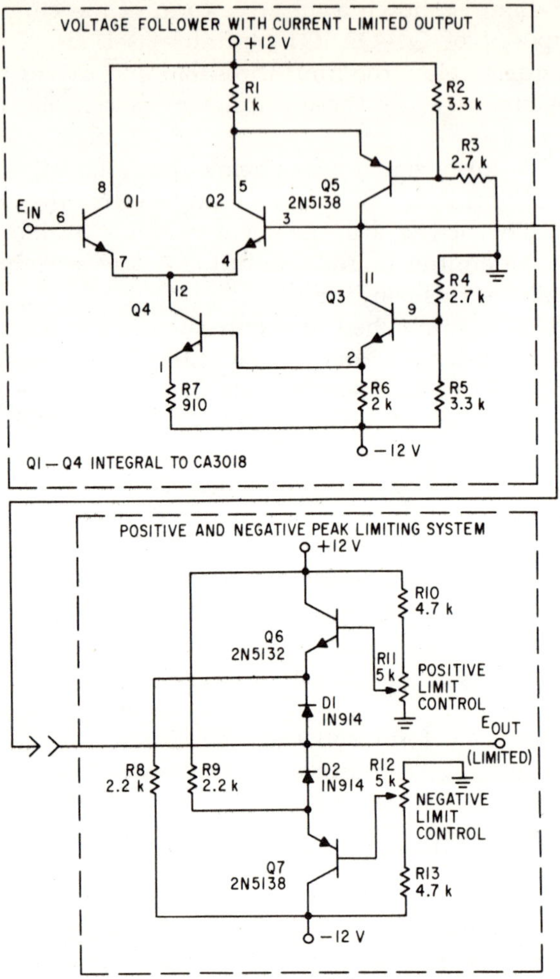

Fig. 8-14 *Zero-offset voltage follower with peak limiting*

The circuit features wide bandwidth because of the common/collector–collector/base drive of the Q1–Q2 combination and the absence of Miller effect. Other desirable characteristics include a ±5-V common-mode range and a predictable current-limited output caused by Q5 and Q3 current-source biasing. Rise and fall times are on the order of 40 ns, but the actual bandwidth realized will be dependent upon the load capacitance seen. This should be minimized, of course, for best bandwidth results.

By adding a second section to the circuit, the output voltage swing can be clipped or limited to a desired level. Introducing biased diodes

D1 and D2 allows the output current to be absorbed above or below a preset voltage level. Thus, the output becomes peak limited. Because of the bidirectional nature of the circuit, the clipping is easily applied to both negative and positive peaks. The clipping potentials can be made variable, as shown with controls R11 and R12.

Bidirectional Current-Mode Limiter

A very interesting, yet simple, form of limiter can be constructed with a diode bridge such as a 3019, as shown in Fig. 8-15. In this circuit, R1 and R2 set up equal dc currents, which flow in diodes D2–D4 and D1–D3. The magnitude of these currents determines the peak output level that will be seen across R3, the output load resistor.

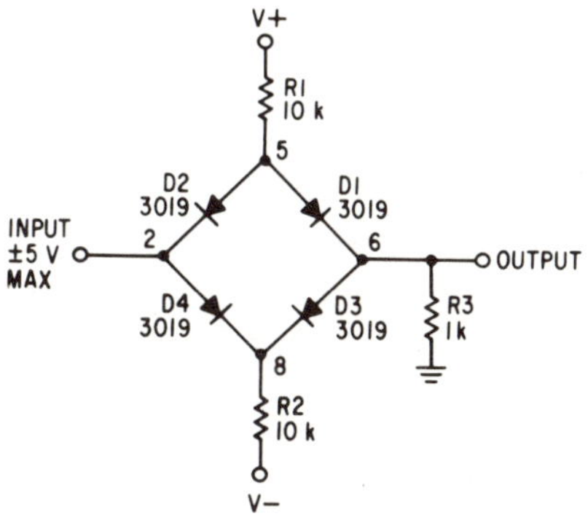

Fig. 8-15 *Bidirectional current-mode limiter*

Ideally, the currents are set up by actual current sources, which will allow high output voltages with low distortion. However, for low output voltages and high supply voltages, resistive "current sources" such as R1–R2 will suffice. Note that the lower the value of R3 relative to R1–R2, the more R1 and R2 act as current sources.

Limiting in this circuit is abrupt because, as the limit point is reached, D2 (or D4) turns off and D1 (or D3) is on. Note that input levels must be confined to levels no more than 5 V greater than the output to prevent Zener breakdown of the bridge.

DC Restorer

All previously described circuits that limit or clip signal levels and peaks assume as an input prerequisite a stable dc baseline. However, it is sometimes necessary in ac-coupled circuits (such as TV waveform processing, for instance) to dc-restore (or clamp) a waveform to a dc reference level.

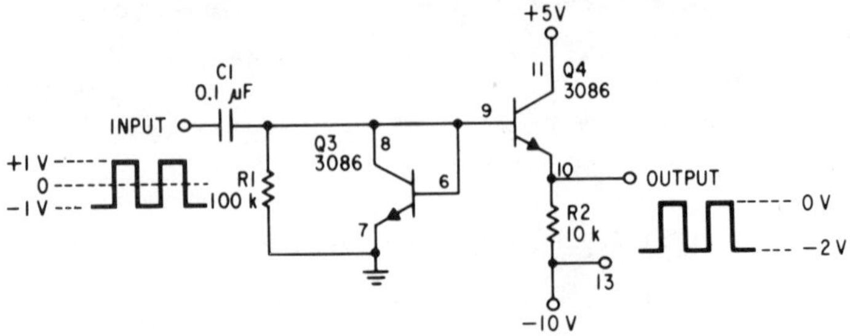

Fig. 8-16 *DC restorer*

The circuit of Fig. 8-16 is very useful as a dc restorer, being able to clamp the most positive peak of an input waveform to ground. C1 and R1 provide input RC coupling to Q3, which is connected as a diode. The positive peaks are clamped by Q3 to a +1 V_{BE} dc reference. Q4 is a matched device from a 3086 array that provides a compensating level shift of −1 V_{BE} plus the buffering of the emitter-follower connection. The output waveform is then dc-restored to 0 V on positive peaks.

Negative-peak dc restoration can be accomplished by reversing Q3 and providing a pnp buffer in place of Q4 (such as from a 3096 device). If a low-impedance variable dc source is provided for the emitter of Q3, the dc restoration can be made to a variable reference level.[8]

Zero-Crossing Detectors

A zero-crossing detector is a specialized form of comparator that is used to extract the zero axis crossing information from an ac signal for timing purposes. Performance parameters important to this application are wide dynamic range, fast response, and low offset, plus, of course, an output compatible with logic. This section discusses several zero-crossing detectors that may use IC arrays.

Figure 8-17, a simple example of such a circuit, is suitable for audio-frequency uses.[9] This circuit uses the 3018 array as a single-stage op amp, with a TTL/compatible output drive. Q1–Q2 is a conventional differential input pair for this op amp, with Q2 loaded by R6 and Q3 serving as an output emitter follower. To allow for high gain, R6 is bootstrapped to zener-connected Q4, providing greater than 50 dB of voltage gain from only the single stage of amplification. The gain is desired to be high so as to maximize the working dynamic range.

As can be noted from the circuit hookup, both positive and negative feedback are used around the Q1–Q4 operational amplifier. A negative feedback path passes through diodes D1–D4 and limits the output

---

[8] See Ref. (17) for a circuit that accomplishes this.

[9] For source, see W. G. Jung, Ref. (16).

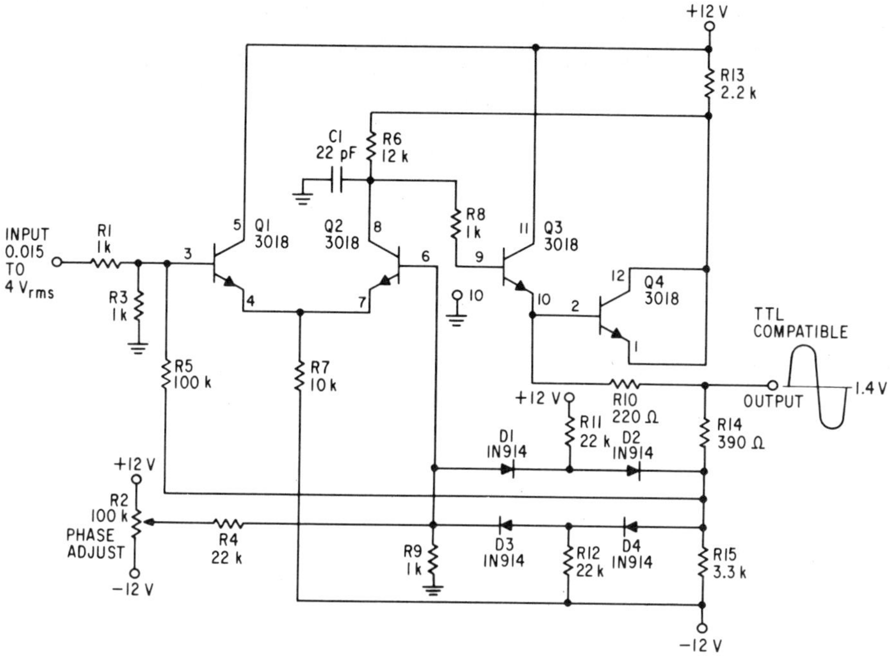

Fig. 8-17 *Simple zero-crossing detector*

peaks to ±2 diode drops. When this swing is centered about the TTL transition region of +1.4 V, a TTL drive source is provided. As a result of the logarithmic nature of the output voltage swing with increasing signal input, a wide range of input variation will trigger a TTL stage connected to the output. With circuit values shown, this range is on the order of 50 dB, from 4 V (max) down to 15 mV. A positive feedback path through R5 gives the overall circuit a defined lower threshold and provides a sharp transition, thereby producing the desirable advantage of allowing noise-free zero axis detection.

An effective means of phase control of the zero axis crossing is provided by pot R2, which offsets the voltage center of the input transition by varying bias on Q2. Since this control is not necessary to the circuit's basic operation, it can be considered optional.

The circuit can be modified in several ways for greater utility. Greater input amplitudes are possible simply by increasing R1. Although virtually any (greater) amplitude can thus be accommodated, the high-to-low operating dynamic range will still be 50 dB.

The circuit as shown has a *noninverting* polarity characteristic, input-to-output. An inverting type of operation can be realized by lifting the ground end of R9 and applying the signal through R9 to pin 6 of the 3018. Finally, the set of discrete IN914 diodes might more efficiently consist of a 3019 or 3039 diode array.

Somewhat more flexibility can be achieved in a zero crossing detector using an op amp if an IC array is used to shape the feedback characteristics only and a standard op amp to provide the working gain. A circuit of this type is shown in Fig. 8-18.

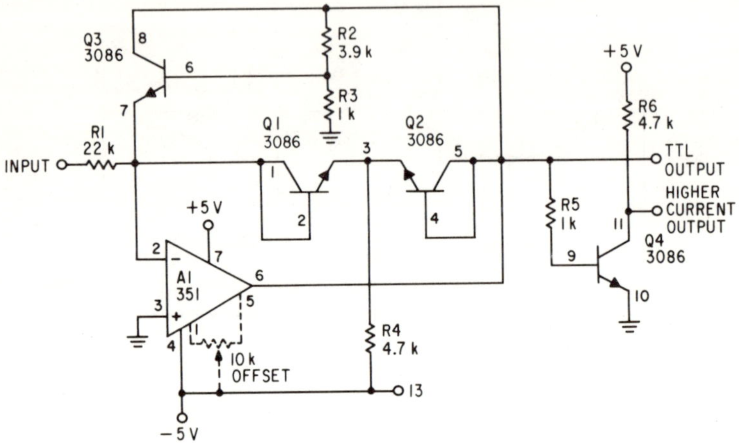

Fig. 8-18 *Zero-crossing detector with an op amp*

In this circuit, an FET input op amp with both low input current and low offset voltage is used for the amplifier, whereas a 3086 array is used to provide the feedback. The negative feedback is again of a logarithmic nature, which allows a wide input dynamic range with relatively small output changes. A basic difference here, however, is that the output voltage is shaped to be *asymmetrical* so as to provide a TTL (or other logic) compatible output.

The output voltage from A1 is set for positive swings by Q3 and R2–R3. Q3 is a V_{BE} multiplier that returns current to the summing point and in so doing provides a positive output limit voltage equal to (R2 + R3)/R3 times the V_{BE} of Q3.

The negative limit of the output voltage is set by diode-connected Q1–Q2, which clamps the output to a voltage just slightly above ground. These two measures provide an output swing that ranges from 0.2 V to +3 V and can furnish about 5 mA of drive over a very wide input dynamic range.

The actual dynamic range of the circuit realized is limited by several factors. The maximum input voltage is 20 V pp, which defines the upper dynamic range limit. The lower dynamic range limit (at frequencies of 100 Hz or less) is determined by the offset voltage of the op amp. If the amplifier's offset voltage is adjusted to maximize input sensitivity, the useful dynamic range of this circuit is 90 dB or more. This value deteriorates at higher frequencies because of the gain falloff in the op amp, but it is still about 50 dB at 10 kHz. Higher gain-bandwidth op amps such as the 318 can improve this figure.

The above comments apply to the output of an op amp used directly. If Q4 is used as a saturated inverter, as shown, the lower limit of sensitivity is even greater than that quoted. This stage is an option, however, and is not necessary to the basic circuit's function.

High-Speed Zero-Crossing Detector

A high-speed form of comparator that is suitable for use as a zero-crossing detector over a wide range of frequencies and amplitude is shown in Fig. 8-19. This circuit uses two arrays in a nonsaturated mode to avoid storage delays.

Array A of the circuit is a 3096, which does the actual comparison. Q1–Q2 (A) forms an npn input stage that is loaded by a pnp current mirror, Q4–Q5 (A). Q1–Q2, operated as an emitter-coupled pair, do not saturate when overdriven and operate as a current-mode limiter only, thereby retaining maximum speed capability.

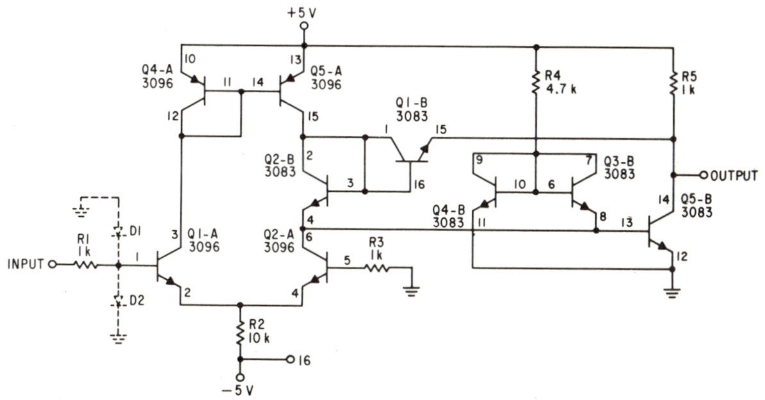

Fig. 8-19 *High-speed zero-crossing detector*

The current mirror performs a differential-to-single-ended conversion, developing a single-ended output to drive array B, a 3083 operated as a 5-V logic output stage. This logic stage is not entirely conventional, however, as it also is operated in a nonsaturated mode.

Q5 (B) of the 3083 is the actual logic stage; it is driven by Q5 (A) and Q2 (A) of the 3096. Forward base current is applied through Q2 (B), and reverse current from Q2 (A). Saturation of Q5 (B) is prevented by diode-connected Q1 (B), and saturation of Q2 (A), by the diode clamp, Q3–Q4 (B). As a result of these measures, no stage in the circuit is ever allowed to saturate and thus lose speed.

This circuit has an input sensitivity that is essentially equal to the offset voltage of Q1–Q2 (A), and full sensitivity is realized to frequencies as high as 100 kHz. Input current (as shown) is approximately 1 μA. If even greater gain bandwidth is desired, R2 can be lowered (with an increase of input current, however). Input diode clamps such as D1–D2 (shown dotted) must be used for high-level inputs to prevent distraction of the input stage.

Interface Drivers

It is often necessary to interface low-power control circuitry to larger current or higher voltage peripherals, such as LEDs, relays, lamps, or other such devices. The higher current IC arrays make such tasks relatively simple for loads up to 100 mA or more and voltages to 50 V.

LED Driver

One of the most simple and commonly seen tasks is to drive an LED from a logic stage. Methods for achieving this function are shown in Fig. 8-20.

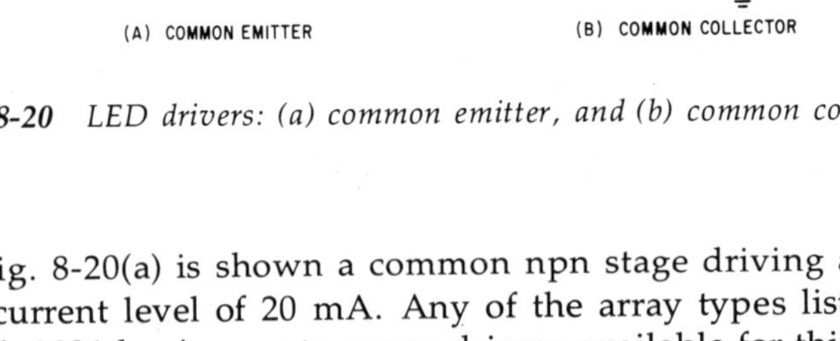

Fig. 8-20 *LED drivers: (a) common emitter, and (b) common collector*

In Fig. 8-20(a) is shown a common npn stage driving a LED to a (typical) current level of 20 mA. Any of the array types listed can be used, with 3081 having up to seven drivers available for this function.

Since this is a saturated stage, the input resistor R1 must provide adequate base drive to Q1 for the HIGH logic level used. For a 20-mA output, this should be 1 to 2 mA to ensure saturation of these transistor types. Suitable resistor values for 5-V (TTL) and 10- to 15-V CMOS logic sources are listed.

The common collector driver in Fig. 8-20(b) is not a saturated driver, since R1 is tied to the same potential as the collector of Q1. R1 is a pullup resistor to ensure a full 5-V swing from a TTL source and thus maximum current in the LED. If a 5-V CMOS stage is used as a driver, R1 is unnecessary since CMOS swings a full 5 V.

50-mA Driver

Some general-purpose, medium-current drivers are shown in Fig. 8-21. They can sink currents of up to 50 mA in their saturated state. They can use 3081, 3083, or 3093 transistors up to 15 V, or 3183 units for up to 50 V.

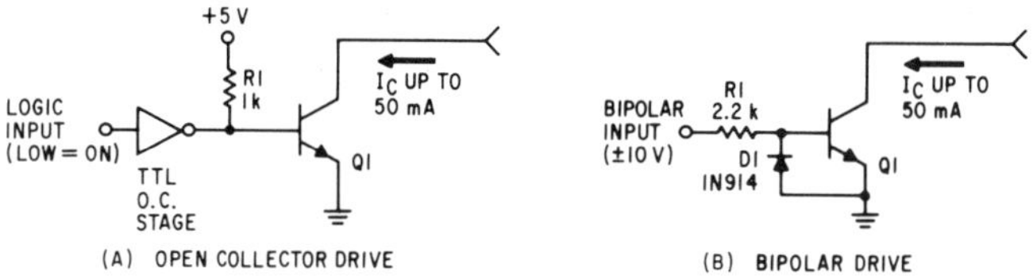

Fig. 8-21 *50-mA saturated drivers: (a) open-collector drive, and (b) bipolar drive*

The circuit in Fig. 8-21(a) is well suited for being driven from an open collector npn stage such as TTL (or one of the OC logic stages of Chap. 8). R1 sets the base drive to Q1, since it drops the full supply voltage and is selected here for 5 mA to provide a 50-mA saturated sink current in Q1. For less current in Q1, R1 can be reduced proportionally. Higher currents in Q1 can be provided by paralleling devices and increasing base drive in proportion. A 3081 can sink up to 350 mA in this fashion, for example.

The stage of Fig. 8-21(b) is similar in terms of its current sink capability and can also be parallel-operated for higher currents. The base drive to Q1 is set by R1, and also the level of the positive input swing. Here, R1 is suitable for a 10-V (or more) swing such as from a comparator or op amp output. D1 should be used if the input can swing more than 5 V negative.

Relay Driver

Either of the stages just described could be used as a relay driver for 5-V or 12-V coil relays by using a corresponding supply voltage. When driving relays, a reverse clamp diode should *always* be used across the coil, as shown in Fig. 8-22. As this circuit shows, 24-V relays can be driven from a 3183 driver, with suitable base drive provided by R1 for a TTL source.

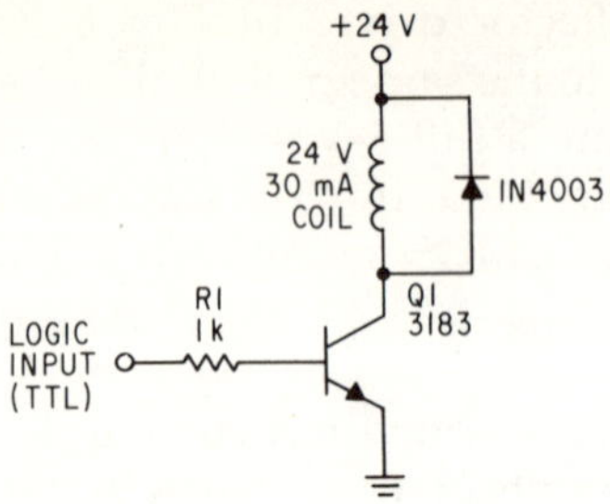

Fig. 8-22 Relay driver

Lamp Driver

Incandescent lamps of suitable current rating can also be driven, with precautions, by 3183 (or other) array transistors, as shown in Fig. 8-23.

A problem that arises with incandescent lamps is their very high current surge when turning on—a result of their nonlinear current/voltage characteristic. This surge can be 10 or more times the final (rated) current, and in some cases it can blow out the driver transistor. A popular solution is to provide a small "keep alive" current in the lamp when off so as to bias it to a small (dim) current level. The surge is then less drastic in its effect on the driver. Resistor R2 should be selected to suit the particular lamp in use, of course.

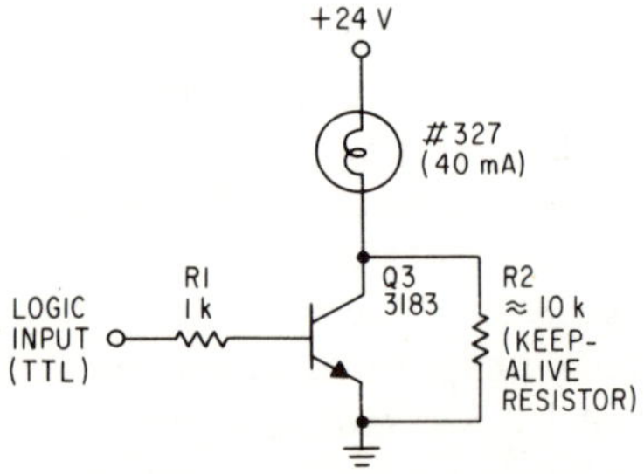

Fig. 8-23 Lamp driver

Noise Generator

A source of wideband noise is often required, particular in signal-synthesis applications such as electronic music. A simple solution to this problem using a 3018 array is shown in Fig. 8-24.

It is well known that Zener diodes generate appreciable noise when biased near their knee or threshold. Since Zener diodes as well as

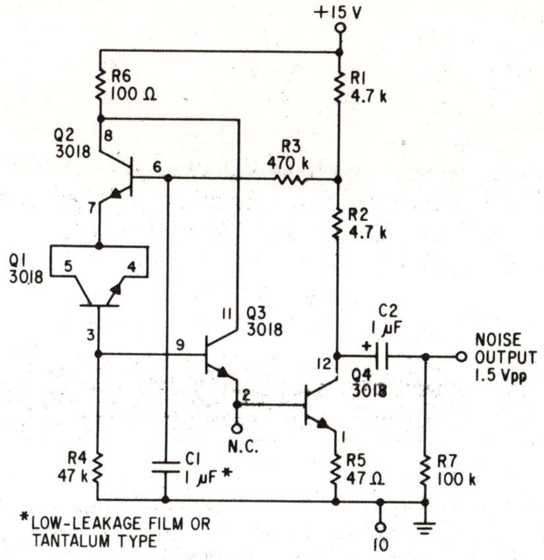

Fig. 8-24 Noise generator

npn transistors are readily available in common arrays, it is relatively simple to exploit this property.

In Fig. 8-24, Q1 of a 3018 array is connected as a Zener in a dc feedback loop around common-emitter amplifier Q4. With this loop operating, the base of Q3 will bias to about 1.5 V. Since this voltage is across R4, the value of R4 sets the current in Zener Q1; here, it is 30 μA for optimum noise generation. Although the feedback path from Q4 through R3 and Q2 to the Zener provides dc coupling, ac feedback is eliminated by bypass C1. This allows high ac gain to realized in Q4.

With the values shown, noise output is 1.5 V pp. Since output impedance of the circuit is 10k, it should feed loads of 100k or more to avoid reduction in output. Also, capacitive loading should be minimized to minimize reduction of noise bandwidth.

No components are critical, except perhaps C1, which should be a low-leakage film or tantalum unit to avoid upsetting dc bias.

REFERENCES FOR CHAPTER 8

(1) ———, "Linear Integrated Circuits," RCA Series IC-47, 1970.

(2) Grebene, A. B., *Analog Integrated Circuit Design,* Van Nostrand Reinhold, N.Y., 1972.

(3) Gilbert, B., "A Precise Four-Quadrant Multiplier with Sub-Nanosecond Response," *IEEE J. Solid State Circuits,* Dec. 1968.

(4) Sheingold, D., (ed), *Non Linear Circuits Handbook,* Analog Devices, Norwood, Mass., 1974.

(5) Jung, W. G., *IC Op Amp Cookbook,* Howard W. Sams, Indianapolis, Ind., 1974.

(6) Jung, W. G., "Get Gain Control of 80 to 100 dB," *Electronic Design,* June 21, 1974.

(7) Jung, W. G., "Application of the Two Quadrant Transconductance Amplifier/Multiplier In Audio Signal Processing," *Journal of the AES,* Apr. 1975.

(8) Jung, W. G., and Todd, C. C., "Operation and Uses for the 570/571 IC Compandor Chip," *The Audio Amateur,* Issue 4, 1976.

(9) Todd, C. C., and Stellrecht, H., "A Monolithic Analog Compandor," *IEEE JSSC,* Dec. 1976.

(10) Jung, W. G., "Gain Control IC for Audio Signal Processing," *Ham Radio,* July 1977.

(11) Jung, W. G., "The Monolithic Balanced Modulator As a Versatile Audio Switching Element," presented at 41st AES convention, Oct. 1971 (AES preprint 816).

(12) Jung, W. G., "Automating the Audio Control Function," *db,* Apr., May, Aug./Sept. 1972.

(13) Theriault, G. E., and Tipping, R., "Application of the CA3019 Diode Array," RCA ICAN-5299, Nov. 1966.

(14) Jung, W. G., "A Survey of Application Ideas for IC's," *Broadcast Engineering* Sept., Oct., Dec. 1969.

(15) Jung, W. G., "Voltage Follower Has Separate Limiting for Each Polarity," *Electronic Design,* Oct. 25, 1970.

(16) Jung, W. G., "Logarithmic-Feedback Zero-Crossing Detector," *Electronic Design,* Nov. 9, 1972.

(17) Jung, W. G., "Temperature Compensated DC Restorer Offers Variable Clamping," *Electronic Design,* Feb. 4, 1971.

(18) Jung, W. G., "IC Lends Stability to Video Limiter," *EDN,* Dec. 15, 1969.

(19) Theriault, G. E., Leidech, A. J., and Campbell, T. H., "Application of the CA3018 IC Transistor Array," RCA ICAN-5296, Oct. 1967.

(20) Miller, E. M., "Rectifier Circuits for Linear ICs," Monochip APN-6, Interdesign.

(21) Data Sheets: MC1445, MC 1496 (Motorola) and NE570/571 (Signetics).

Appendix A

Selected Manufacturer's Data Sheets

This appendix contains specification sheets for the more popular IC array devices. The data sheets are reproduced here through the courtesy of the original manufacturer of each device. In most cases, the devices are also available from multiple sources.

It is recommended that the reader avail himself of complete (and the most current) information on IC array devices by writing the manufacturers directly. Manufacturer's addresses are given in Appendix B.

RCA
Solid State
Division

Linear Integrated Circuits

Monolithic Silicon

CA3045, CA3045F, CA3045H, CA3045L, CA3046

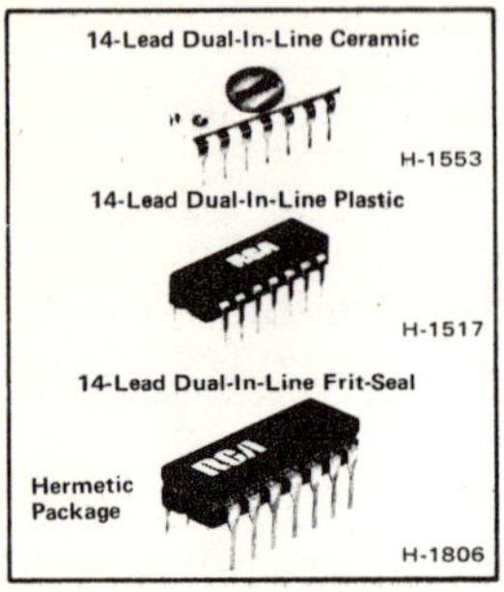

General-Purpose Transistor Arrays

THREE ISOLATED TRANSISTORS AND ONE DIFFERENTIALLY-CONNECTED TRANSISTOR PAIR

For Low-Power Applications at Frequencies from DC through the VHF Range

Features

- Two matched pairs of transistors
  V_{BE} matched ± 5 mV
  Input offset current 2 μA max. at I_C = 1 mA
- 5 general purpose monolithic transistors
- Operation from DC to 120 MHz
- Wide operating current range
- Low noise figure - - 3.2 dB typ. at 1 kHz
- Full military temperature range for CA3045
  -55 to +125°C

The CA3045 and CA3046 each consist of five general-purpose silicon n-p-n transistors on a common monolithic substrate. Two of the transistors are internally connected to form a differentially-connected pair.

The transistors of the CA3045 and CA3046 are well suited to a wide variety of applications in low power systems in the DC through VHF range. They may be used as discrete transistors in conventional circuits. However, in addition, they provide the very significant inherent integrated circuit advantages of close electrical and thermal matching.

The CA3045 is supplied in a 14-lead dual-in-line hermetic (welded-seal) ceramic package and the CA3045F in a 14-lead dual-in-line hermetic (frit-seal) ceramic package.

The CA3046 is electrically identical to the CA3045 but is supplied in a dual-in-line plastic package for applications requiring only a limited temperature range.

Applications

- General use in all types of signal processing systems operating anywhere in the frequency range from DC to VHF
- Custom designed differential amplifiers
- Temperature compensated amplifiers
- See RCA Application Note, ICAN-5296 "Application of the RCA-CA3018 Integrated-Circuit Transistor Array" for suggested applications.

The CA3045 and CA3046 are available in the packages shown below

Package	Suffix Letter	CA3045	CA3046
14-Lead Dual-In-Line Plastic	E		√
14-Lead Dual-In-Line Ceramic	D	√	
14-Line Dual-In-Line Frit-Seal Ceramic	F	√	
Beam Lead	L	√	
Chip	H	√	

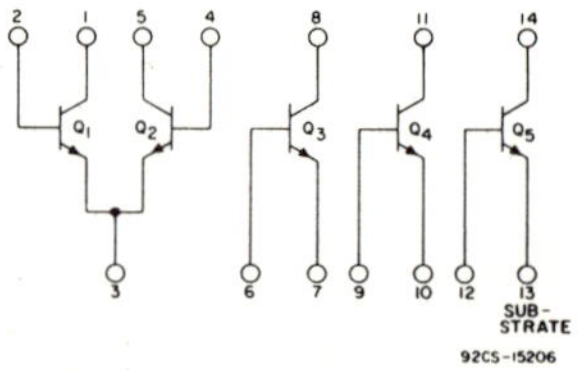

Fig.1 - Schematic diagram.

Printed in USA/11-73

Supersedes issue dated 3-70

File No.341 CA3045, CA3046

ABSOLUTE MAXIMUM RATINGS AT T_A = 25°C

	CA3045 Each Transistor	CA3045 Total Package	CA3046, CA3045F Each Transistor	CA3046, CA3045F Total Package	
Power Dissipation:					
T_A up to 55°C	–	–	300	750	mW
T_A > 55°C	–	–	Derate at 6.67		mW/°C
T_A up to 75°C	300	750	–	–	mW
T_A > 75°C	Derate at 8		–	–	mW/°C
Collector-to-Emitter Voltage, V_{CEO}	15	–	15	–	V
Collector-to-Base Voltage, V_{CBO}	20	–	20	–	V
Collector-to-Substrate Voltage, V_{CIO}*	20	–	20	–	V
Emitter-to-Base Voltage, V_{EBO}	5	–	5	–	V
Collector Current	50	–	50	–	mA
Temperature Range:					
Operating	−55 to +125		−55 to +125		°C
Storage	−65 to +150		−65 to +150		°C
Lead Temperature (During Soldering): At distance 1/16 ±1/32" (1.59 ±0.79 mm) from case for 10 seconds max.	+265		+265		°C

* The collector of each transistor of the CA3045 and CA3046 is isolated from the substrate by an integral diode. *The substrate (terminal 13) must be connected to the most negative point in the external circuit to maintain isolation between transistors and to provide for normal transistor action.*

ELECTRICAL CHARACTERISTICS, at T_A = 25°C

Characteristics apply for each transistor in the CA3045 and CA3046 as specified.

CHARACTERISTICS	SYMBOLS	SPECIAL TEST CONDITIONS	LIMITS Type CA3045 Type CA3046 MIN.	TYP.	MAX.	UNITS	CHARAC-TERISTIC CURVES FIG.
STATIC CHARACTERISTICS							
Collector-to-Base Breakdown Voltage	$V_{(BR)CBO}$	I_C = 10 μA, I_E = 0	20	60	-	V	-
Collector-to-Emitter Breakdown Voltage	$V_{(BR)CEO}$	I_C = 1 mA, I_B = 0	15	24	-	V	-
Collector-to-Substrate Breakdown Voltage	$V_{(BR)CIO}$	I_C = 10 μA, I_{CI} = 0	20	60	-	V	-
Emitter-to-Base Breakdown Voltage	$V_{(BR)EBO}$	I_E = 10 μA, I_C = 0	5	7	-	V	-
Collector-Cutoff Current	I_{CBO}	V_{CB} = 10 V, I_E = 0	-	0.002	40	nA	2
Collector-Cutoff Current	I_{CEO}	V_{CE} = 10 V, I_B = 0	-	See curve	0.5	μA	3
Static Forward Current-Transfer Ratio (Static Beta)	h_{FE}	V_{CE} = 3 V, I_C = 10 mA	-	100	-	-	4
		V_{CE} = 3 V, I_C = 1 mA	40	100	-	-	
		V_{CE} = 3 V, I_C = 10 μA	-	54	-	-	
Input Offset Current for Matched Pair Q_1 and Q_2. $\lvert I_{IO_1} - I_{IO_2}\rvert$		V_{CE} = 3 V, I_C = 1 mA	-	0.3	2	μA	5
Base-to-Emitter Voltage	V_{BE}	V_{CE} = 3 V, I_E = 1 mA	-	0.715	-	V	6
		V_{CE} = 3 V, I_E = 10 mA	-	0.800	-		
Magnitude of Input Offset Voltage for Differential Pair $\lvert V_{BE_1} - V_{BE_2}\rvert$		V_{CE} = 3 V, I_C = 1 mA	-	0.45	5	mV	6,8
Magnitude of Input Offset Voltage for Isolated Transistors $\lvert V_{BE_3} - V_{BE_4}\rvert$, $\lvert V_{BE_4} - V_{BE_5}\rvert$, $\lvert V_{BE_5} - V_{BE_3}\rvert$		V_{CE} = 3 V, I_C = 1 mA	-	0.45	5	mV	6,8
Temperature Coefficient of Base-to-Emitter Voltage	$\frac{\Delta V_{BE}}{\Delta T}$	V_{CE} = 3 V, I_C = 1 mA	-	−1.9	-	mV/°C	7
Collector-to-Emitter Saturation Voltage	V_{CES}	I_B = 1 mA, I_C = 10 mA	-	0.23	-	V	-
Temperature Coefficient: Magnitude of Input-Offset Voltage	$\frac{\lvert \Delta V_{IO}\rvert}{\Delta T}$	V_{CE} = 3 V, I_C = 1 mA	-	1.1	-	μV/°C	8

ELECTRICAL CHARACTERISTICS (Cont'd.)

DYNAMIC CHARACTERISTICS							
Low-Frequency Noise Figure	NF	$f = 1$ kHz, $V_{CE} = 3$ V, $I_C = 100\ \mu A$ Source Resistance = 1 kΩ	-	3.25	-	dB	9(b)
Low-Frequency, Small-Signal Equivalent-Circuit Characteristics:							
Forward Current-Transfer Ratio	h_{fe}	$f = 1$ kHz, $V_{CE} = 3$ V, $I_C = 1$ mA	-	110	-	-	10
Short-Circuit Input Impedance	h_{ie}		-	3.5	-	kΩ	
Open-Circuit Output Impedance	h_{oe}		-	15.6	-	µmho	
Open-Circuit Reverse Voltage-Transfer Ratio	h_{re}		-	1.8×10^{-4}	-	-	
Admittance Characteristics:							
Forward Transfer Admittance	Y_{fe}	$f = 1$ MHz, $V_{CE} = 3$ V, $I_C = 1$ mA	-	31-j1.5	-	-	11
Input Admittance	Y_{ie}		-	0.3+j0.04	-	-	12
Output Admittance	Y_{oe}		-	0.001+j0.03	-	-	13
Reverse Transfer Admittance	Y_{re}		-	See curve	-	-	14
Gain-Bandwidth Product	f_T	$V_{CE} = 3$ V, $I_C = 3$ mA	300	550	-	-	15
Emitter-to-Base Capacitance	C_{EB}	$V_{EB} = 3$ V, $I_E = 0$	-	0.6	-	pF	-
Collector-to-Base Capacitance	C_{CB}	$V_{CB} = 3$ V, $I_C = 0$	-	0.58	-	pF	-
Collector-to-Substrate Capacitance	C_{CI}	$V_{CS} = 3$ V, $I_C = 0$	-	2.8	-	pF	-

STATIC CHARACTERISTICS

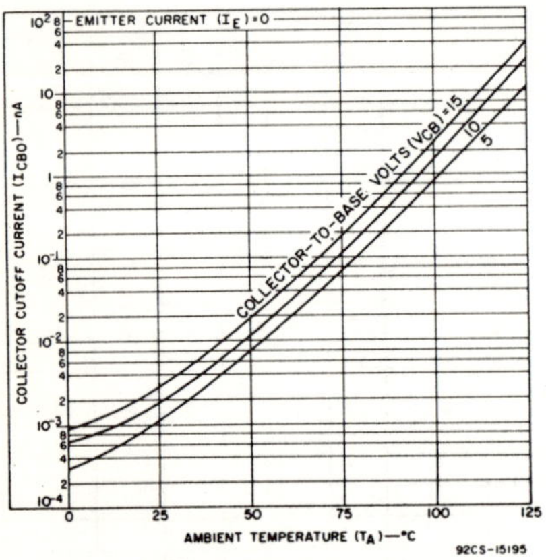

Fig.2 - Typical collector-to-base cutoff current vs ambient temperature for each transistor.

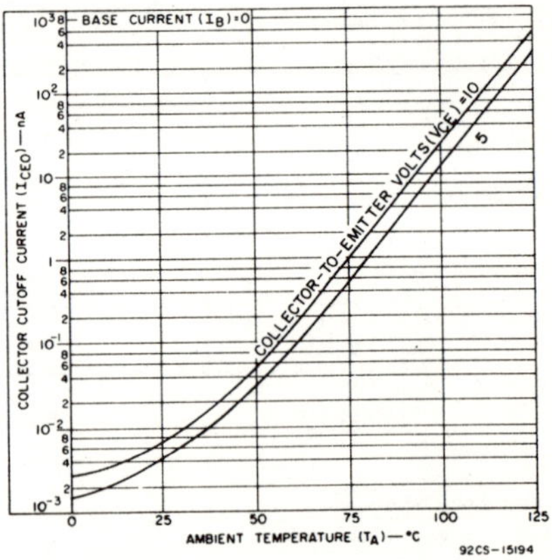

Fig.3 - Typical collector-to-emitter cutoff current vs ambient temperature for each transistor.

STATIC CHARACTERISTICS

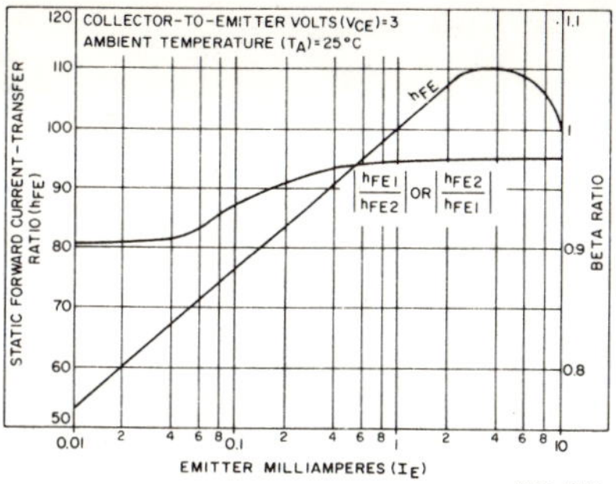

Fig.4 - Typical static forward current-transfer ratio and beta ratio for transistors Q_1 and Q_2 vs emitter current.

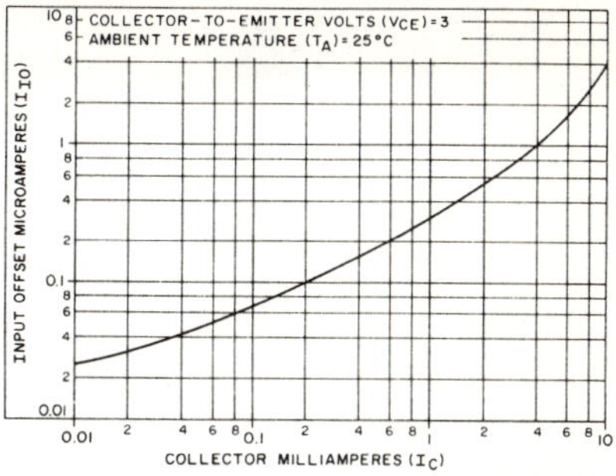

Fig.5 - Typical input offset current for matched transistor pair Q_1Q_2 vs collector current.

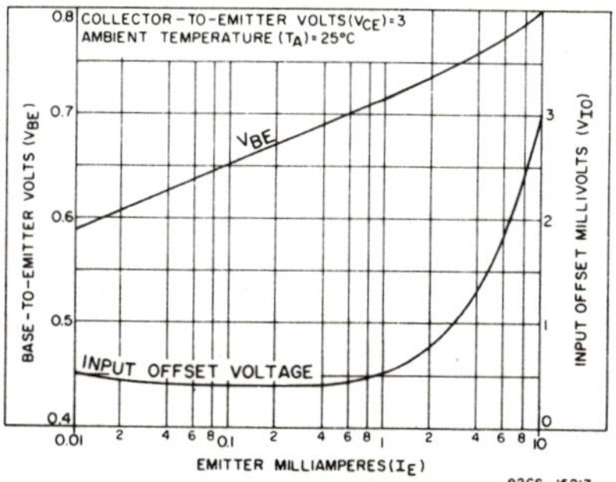

Fig.6 - Typical static base-to-emitter voltage characteristic and input offset voltage for differential pair and paired isolated transistors vs emitter current.

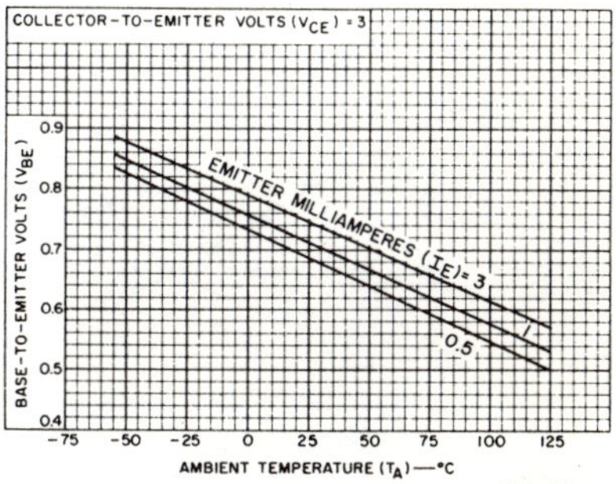

Fig.7 - Typical base-to-emitter voltage characteristic vs ambient temperature for each transistor.

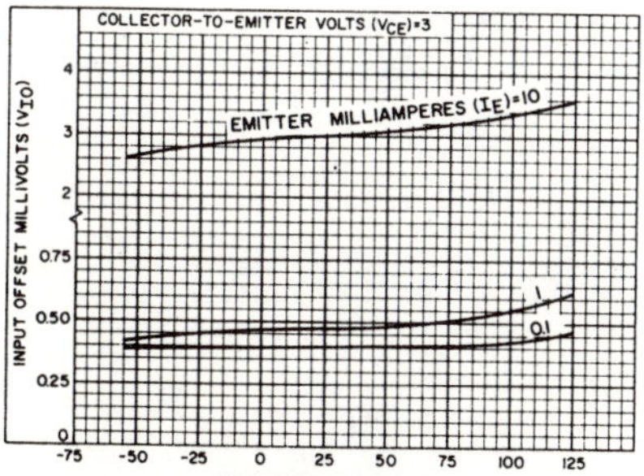

Fig.8 - Typical input offset voltage characteristics for differential pair and paired isolated transistors vs ambient temperature.

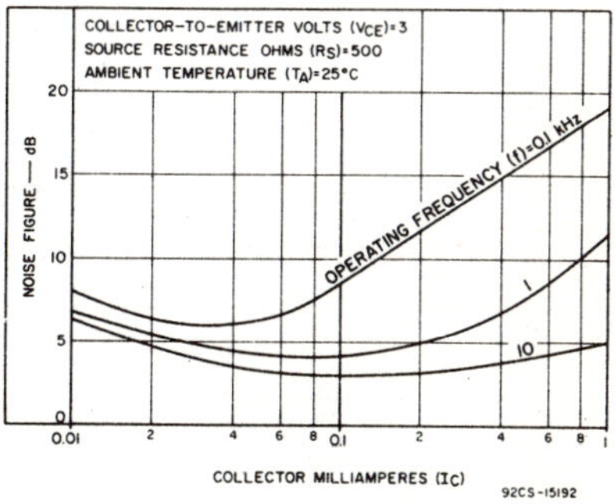

Fig.9(a) - Typical noise figure vs collector current.

DYNAMIC CHARACTERISTICS FOR EACH TRANSISTOR

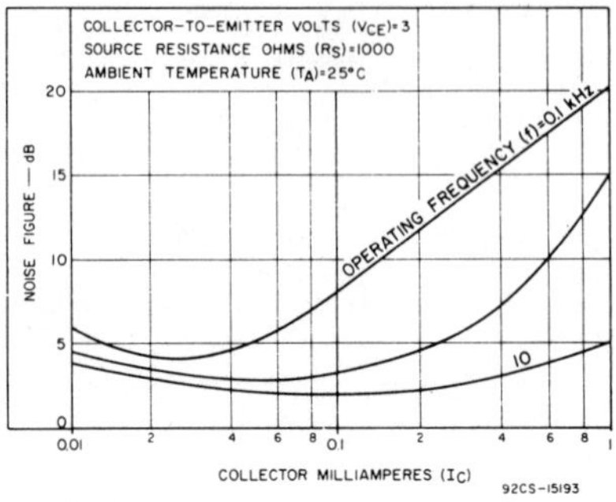

Fig.9(b) - Typical noise figure vs collector current.

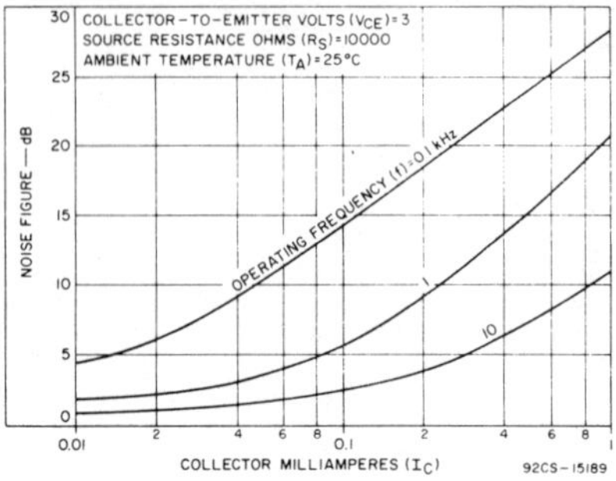

Fig.9(c) - Typical noise figure vs collector current.

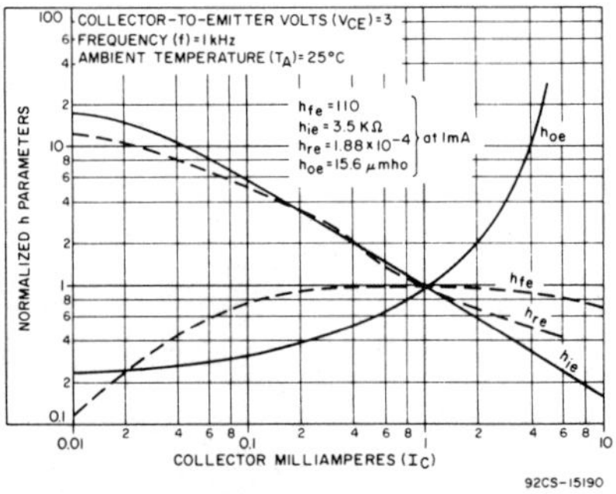

Fig.10 - Typical normalized forward current-transfer ratio, short-circuit input impedance, open-circuit output impedance, and open-circuit reverse voltage-transfer ratio vs collector current.

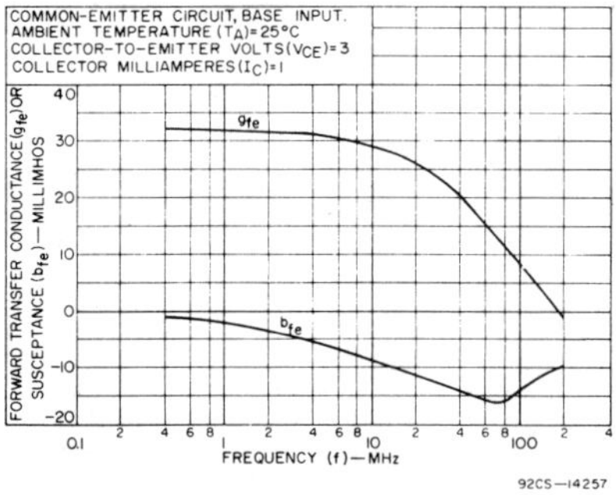

Fig.11 - Typical forward transfer admittance vs frequency.

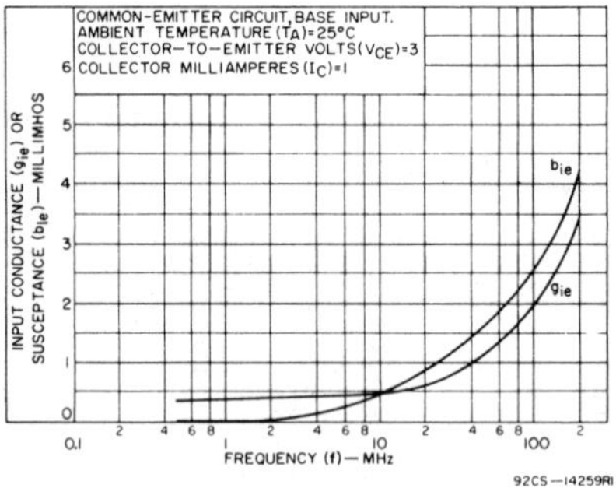

Fig.12 - Typical input admittance vs frequency.

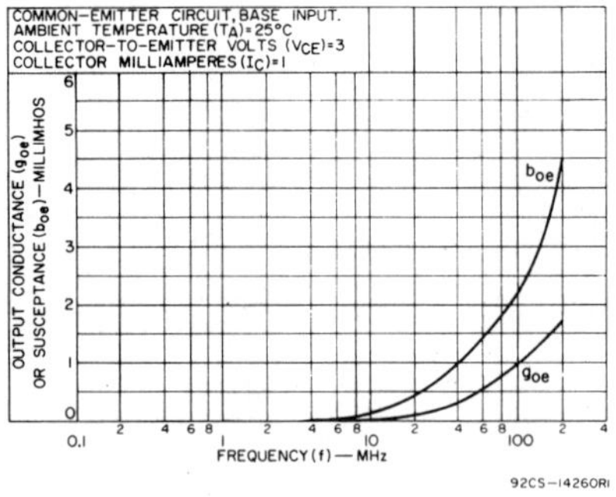

Fig.13 - Typical output admittance vs frequency.

DYNAMIC CHARACTERISTICS FOR EACH TRANSISTOR

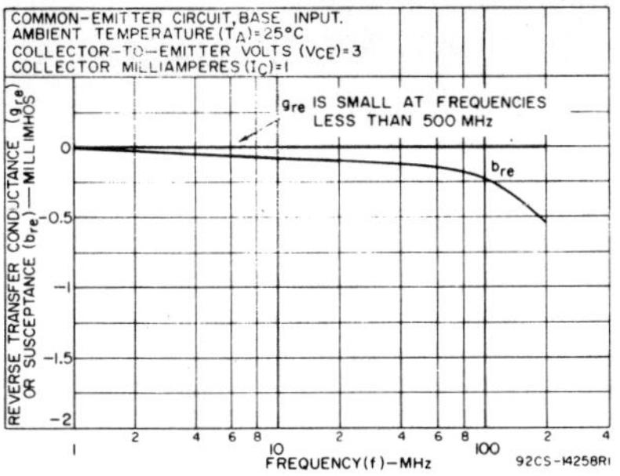

Fig.14 - Typical reverse transfer admittance vs frequency.

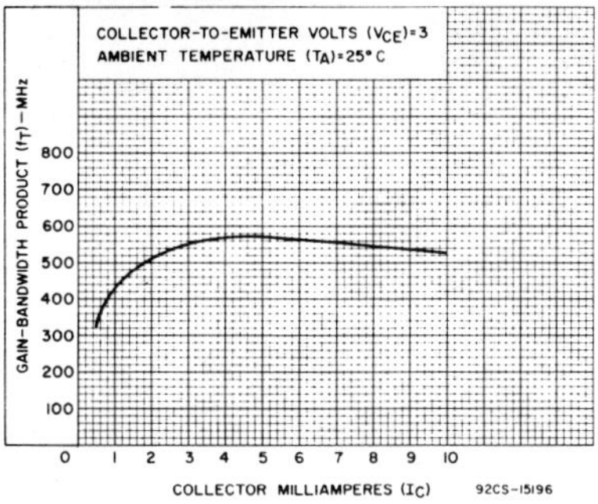

Fig.15 - Typical gain-bandwidth product vs collector current.

DIMENSIONAL OUTLINE CA3045

14-Lead Dual-In-Line

Ceramic Package JEDEC MO-001-AD

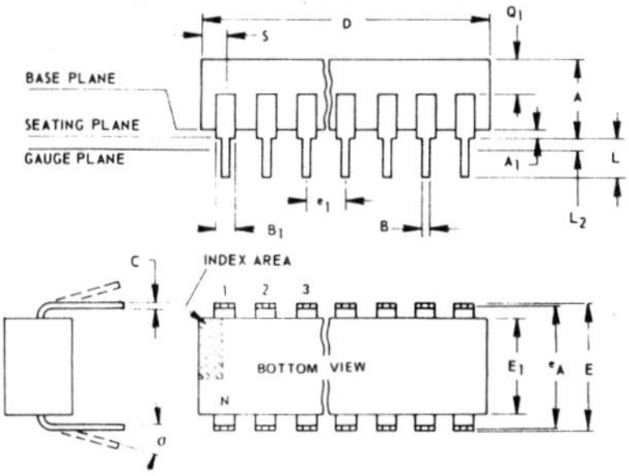

SYMBOL	INCHES		NOTE	MILLIMETERS	
	MIN.	MAX.		MIN.	MAX.
A	0.120	0.160		3.05	4.06
A_1	0.020	0.065		0.51	1.65
B	0.014	0.020		0.356	0.508
B_1	0.050	0.065		1.27	1.65
C	0.008	0.012		0.204	0.304
D	0.745	0.770		18.93	19.55
E	0.300	0.325		7.62	8.25
E_1	0.240	0.260		6.10	6.60
e_1	0.100 TP		2	2.54 TP	
e_A	0.300 TP		2, 3	7.62 TP	
L	0.125	0.150		3.18	3.81
L_2	0.000	0.030		0.000	0.76
α	0°	15°	4	0°	15°
N	14		5	14	
N_1	0		6	0	
Q_1	0.050	0.085		1.27	2.15
S	0.065	0.090		1.66	2.28

92SS-4411R1

NOTES:

1. Refer to Rules for Dimensioning (JEDEC Publication No. 13) for Axial Lead Product Outlines.
2. Leads within 0.005" (0.12 mm) radius of True Position (TP) at guage plane with maximum material condition and unit installed.
3. e_A applies in zone L_2 when unit installed.
4. α applies to spread leads prior to installation.
5. N is the maximum quantity of lead positions.
6. N_1 is the quantity of allowable missing leads.

DIMENSIONAL OUTLINE CA3046

14-Lead Dual-In-Line

Plastic Package and F-Type (Frit-Seal) Ceramic Package

JEDEC MO-001-AB

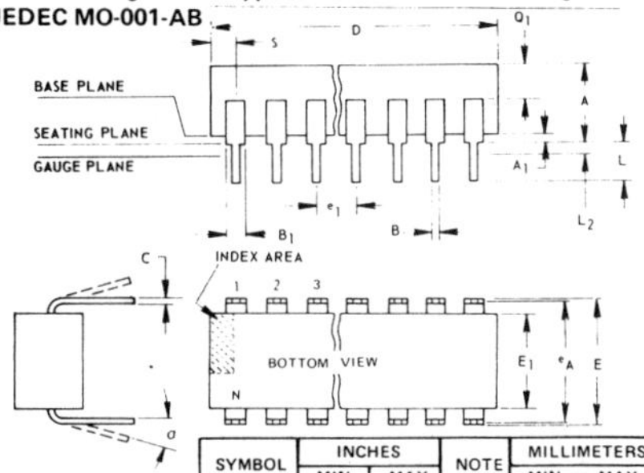

SYMBOL	INCHES		NOTE	MILLIMETERS	
	MIN.	MAX.		MIN.	MAX.
A	0.155	0.200		3.94	5.08
A_1	0.020	0.050		0.51	1.27
B	0.014	0.020		0.356	0.508
B_1	0.050	0.065		1.27	1.65
C	0.008	● 0.012		0.204	0.304
D	0.745	0.770		18.93	19.55
E	0.300	0.325		7.62	8.25
E_1	0.240	0.260		6.10	6.60
e_1	0.100 TP		2	2.54 TP	
e_A	0.300 TP		2, 3	7.62 TP	
L	0.125	0.150		3.18	3.81
L_2	0.000	0.030		0.000	0.76
α	0°	15°	4	0°	15°
N	14		5	14	
N_1	0		6	0	
Q_1	0.040	0.075		1.02	1.90
S	0.065	0.090		1.66	2.28

92SS-4296R1

NOTES:

1. Refer to Rules for Dimensioning (JEDEC Publication No. 13) for Axial Lead Product Outlines.
2. Leads within 0.005" (0.12 mm) radius of True Position (TP) at guage plane with maximum material condition and unit installed.
3. e_A applies in zone L_2 when unit installed.
4. α applies to spread leads prior to installation.
5. N is the maximum quantity of lead positions.
6. N_1 is the quantity of allowable missing leads.

● When this device is supplied solder-dipped, the maximum lead thickness (narrow portion) will not exceed 0.013".

When incorporating RCA Solid State Devices in equipment, it is recommended that the designer refer to "Operating Considerations for RCA Solid State Devices", Form No. 1CE-402, available on request from RCA Solid State Division, Box 3200, Somerville, N.J. 08876.

RCA
Solid State
Division

Linear Integrated Circuits

Monolithic Silicon

CA3083, CA3083F, CA3083H

File Number 481

16-Lead Dual-In-Line
Frit-Seal (Hermetic)
Package H-1807

16-Lead Dual-In-Line
Plastic Package H-1622

General-Purpose High-Current N-P-N Transistor Array

Applications

- Signal processing and switching systems operating from DC to VHF
- Lamp and relay driver
- Differential amplifier
- Temperature-compensated amplifier
- Thyristor firing
- See RCA Application Note, ICAN-5296 "Application of the RCA-CA3018 Circuit Transistor Array" for suggested applications

RCA-CA3083 is a versatile array of five high-current (to 100mA) n-p-n transistors on a common monolithic substrate. In addition, two of these transistors (Q1 and Q2) are matched at low currents (i.e. 1mA) for applications in which offset parameters are of special importance.

Independent connections for each transistor plus a separate terminal for the substrate permit maximum flexibility in circuit design. The CA3083 is supplied in a 16-lead dual-in-line frit-seal ceramic package. The CA3083 is also available in chip form.

Features

- High I_C: 100mA max.
- Low V_{CEsat} (at 50 mA): 0.7V max.
- Matched pair (Q1 and Q2)—
  V_{IO} (V_{BE} matched): ± 5 mV max.
  I_{IO} (at 1 mA): 2.5 μA max.
- 5 independent transistors plus separate substrate connection

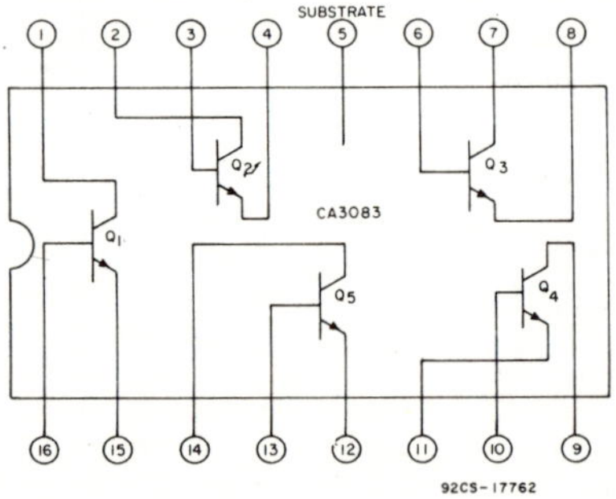

Fig. 1—Functional diagram of the CA3083.

Printed in USA/10-77

Supersedes issue dated 11-73

MAXIMUM RATINGS, Absolute-Maximum Values at T_A = 25°C

Power Dissipation:

Any one transistor	500	mW
Total package	750	mW
Above 55°C	Derate linearly 6.67	mW/°C

Ambient Temperature Range:

Operating	−55 to +125	°C
Storage	−65 to +150	°C

Lead Temperature (During Soldering):

At distance 1/16" ±1/32" (1.59 mm ±0.79 mm) from case for 10 seconds max.	265	°C

The following ratings apply for each transistor in the device:

Collector-to-Emitter Voltage (V_{CEO})	15	V
Collector-to-Base Voltage (V_{CBO})	20	V
Collector-to-Substrate Voltage (V_{CIO})■	20	V
Emitter-to-Base Voltage (V_{EBO})	5	V
Collector Current (I_C)	100	mA
Base Current (I_B)	20	mA

■ The collector of each transistor of the CA3083 is isolated from the substrate by an integral diode. The substrate must be connected to a voltage which is more negative than any collector voltage in order to maintain isolation between transistors and provide normal transistor action. To avoid undesired coupling between transistors, the substrate terminal (5) should be maintained at either DC or signal (AC) ground. A suitable bypass capacitor can be used to establish a signal ground.

ELECTRICAL CHARACTERISTICS at T_A = 25°C
For Equipment Design

CHARACTERISTICS	SYMBOL	TEST CONDITIONS	Typ. Char. Curve Fig. No.	LIMITS Min.	Typ.	Max.	UNITS
For Each Transistor:							
Collector-to-Base Breakdown Voltage	$V_{(BR)CBO}$	I_C = 100μA, I_E = 0	–	20	60	–	V
Collector-to-Emitter Breakdown Voltage	$V_{(BR)CEO}$	I_C = 1mA, I_B = 0	–	15	24	–	V
Collector-to-Substrate Breakdown Voltage	$V_{(BR)CIO}$	I_{CI} = 100μA, I_B = 0, I_E = 0	–	20	60	–	V
Emitter-to-Base Breakdown Voltage	$V_{(BR)EBO}$	I_E = 500μA, I_C = 0	–	5	6.9	–	V
Collector-Cutoff-Current	I_{CEO}	V_{CE} = 10V, I_B = 0	–	–	–	10	μA
Collector-Cutoff-Current	I_{CBO}	V_{CB} = 10V, I_E = 0	–	–	–	1	μA
DC Forward-Current Transfer Ratio	h_{FE}	V_{CE} = 3V, I_C = 10mA	2	40	76	–	
		V_{CE} = 3V, I_C = 50mA		40	75	–	
Base-to-Emitter Voltage	V_{BE}	V_{CE} = 3V, I_C = 10mA	3	0.65	0.74	0.85	V
Collector-to-Emitter Saturation Voltage	V_{CEsat}	I_C = 50mA, I_B = 5mA	4	–	0.40	0.70	V
Gain-Bandwidth Product	f_T	V_{CE} = 3 V, I_C = 10 mA		–	450	–	MHz
For Transistors Q1 and Q2 (As a Differential Amplifier):							
Absolute Input Offset Voltage	$\|V_{IO}\|$	V_{CE} = 3V, I_C = 1mA	7	–	1.2	5	mV
Absolute Input Offset Current	$\|I_{IO}\|$	V_{CE} = 3V, I_C = 1mA	8	–	0.7	2.5	μA

TYPICAL STATIC CHARACTERISTICS FOR EACH TRANSISTOR

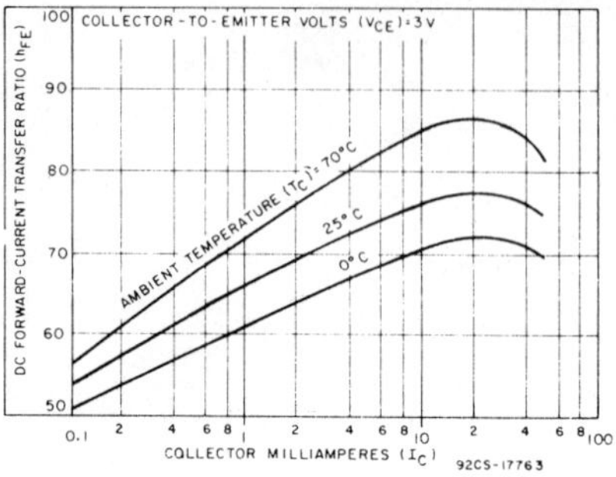

Fig.2 – h_{FE} vs I_C

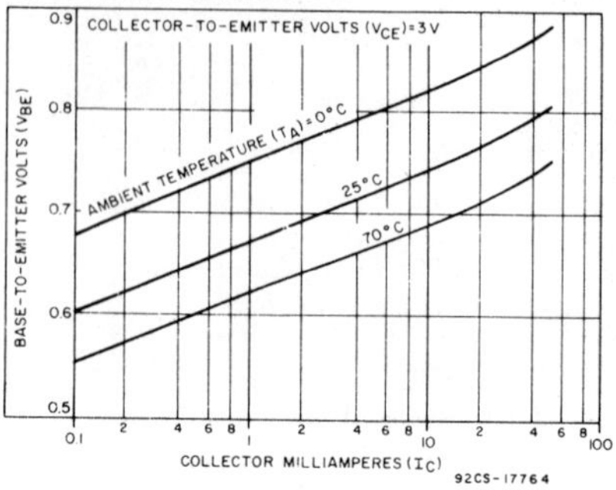

Fig.3 – V_{BE} vs I_C

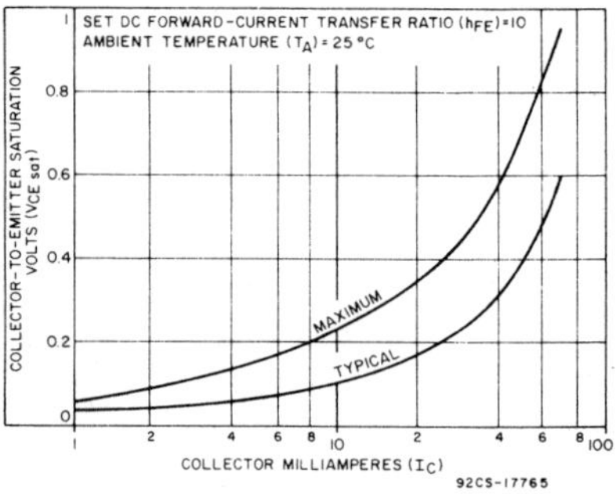

Fig.4 – V_{CEsat} vs I_C at 25°C

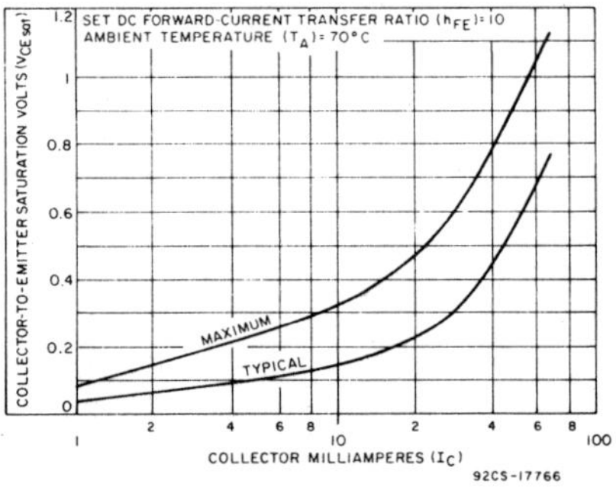

Fig.5 – V_{CEsat} vs I_C at 70°C

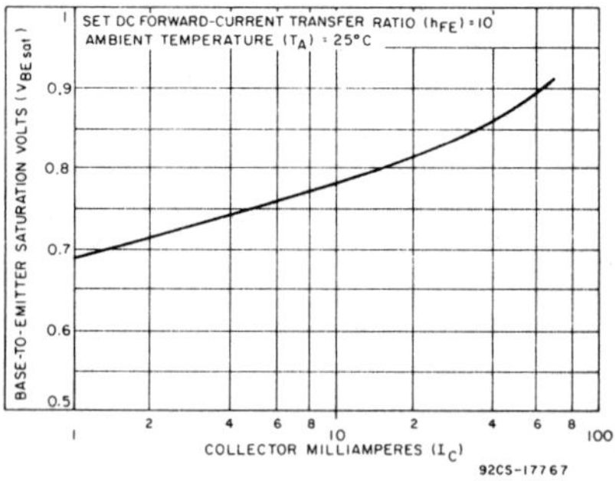

Fig.6 – V_{BEsat} vs I_C

TYPICAL STATIC CHARACTERISTICS FOR DIFFERENTIAL AMPLIFIER

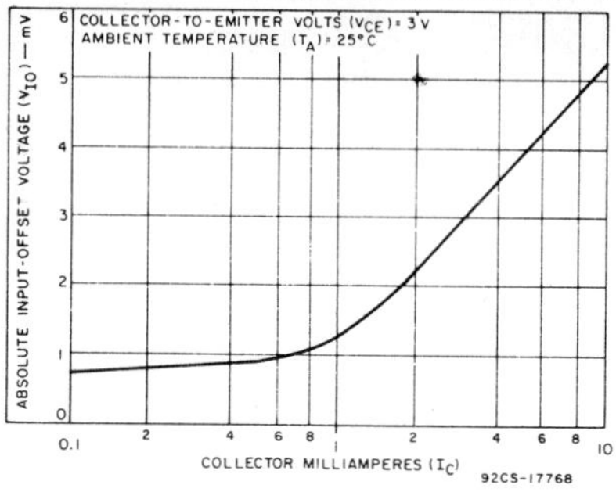

Fig.7 – V_{IO} vs I_C (transistors Q1 and Q2 as a differential amplifier).

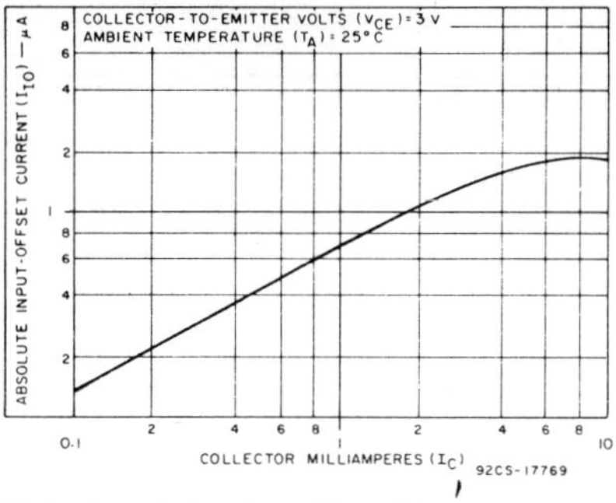

Fig.8 – I_{IO} vs I_C (transistors Q1 and Q2 as a differential amplifier).

DIMENSIONAL OUTLINE

16-LEAD DUAL-IN-LINE PLASTIC OR FRIT-SEAL CERAMIC PACKAGE – JEDEC MO-001-AC

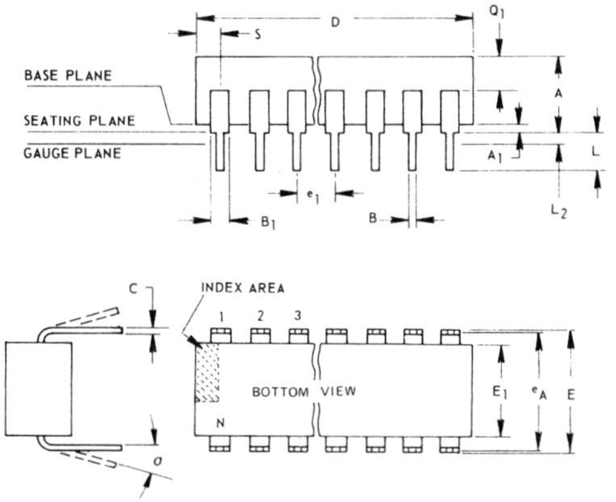

SYMBOL	INCHES		NOTE	MILLIMETERS	
	MIN.	MAX.		MIN.	MAX.
A	0.155	0.200		3.94	5.08
A_1	0.020	0.050		0.51	1.27
B	0.014	0.020		0.356	0.508
B_1	0.035	0.065		0.89	1.65
C	0.008	•0.012		0.204	0.304
D	0.745	0.785		18.93	19.93
E	0.300	0.325		7.62	8.25
E_1	0.240	0.260		6.10	6.60
e_1	0.100 TP		2	2.54 TP	
e_A	0.300 TP		2, 3	7.62 TP	
L	0.125	0.150		3.18	3.81
L_2	0.000	0.030		0.000	0.76
α	0°	15°	4	0°	15°
N	16		5	16	
N_1	0		6	0	
Q_1	0.040	0.075		1.02	1.90
S	0.015	0.060		0.39	1.52

92CM-15967R1

NOTES:

1. Refer to Rules for Dimensioning (JEDEC Publication No. 13) for Axial Lead Product Outlines.
2. Leads within 0.005" (0.12 mm) radius of True Position (TP) at gauge plane with maximum material condition and unit installed.
3. e_A applies in zone L_2 when unit installed.
4. α applies to spread leads prior to installation.
5. N is the maximum quantity of lead positions.
6. N_1 is the quantity of allowable missing leads.

• When this device is supplied solder-dipped, the maximum lead thickness (narrow portion) will not exceed 0.013".

When incorporating RCA Solid State Devices in equipment, it is recommended that the designer refer to "Operating Considerations for RCA Solid State Devices". (Form No. 1CE-402), available on request from RCA Commercial Engineering, Harrison, N. J. 07029.

RCA Solid State Division

Linear Integrated Circuits

Monolithic Silicon

CA3096AE
CA3096E CA3096CE

N-P-N/P-N-P Transistor-Array

16-Lead Dual-in-Line Plastic Package H-1622

Five-Independent Transistors: Three n-p-n and Two p-n-p

Applications:

- Differential Amplifiers
- DC Amplifiers
- Sense Amplifiers
- Level Shifters
- Timers
- Lamp and Relay Drivers
- Thyristor Firing Circuits
- Temperature-Compensated Amplifiers
- Operational Amplifiers

RCA-CA3096CE, CA3096E, and CA3096AE are general-purpose high-voltage silicon transistor arrays. Each array consists of five independent transistors (two p-n-p and three n-p-n types) on a common substrate, which has a separate connection. Independent connections for each transistor permit maximum flexibility in circuit design.

Types CA3096AE, CA3096E, and CA3096CE are identical, except that the CA3096AE specifications include parameter matching and greater stringency in I_{CBO}, I_{CEO}, and $V_{CE}(SAT)$. The CA3096CE is a relaxed version of the CA3096E.

The CA3096CE, CA3096E, and CA3096AE are supplied in 16-lead dual-in-line plastic packages.

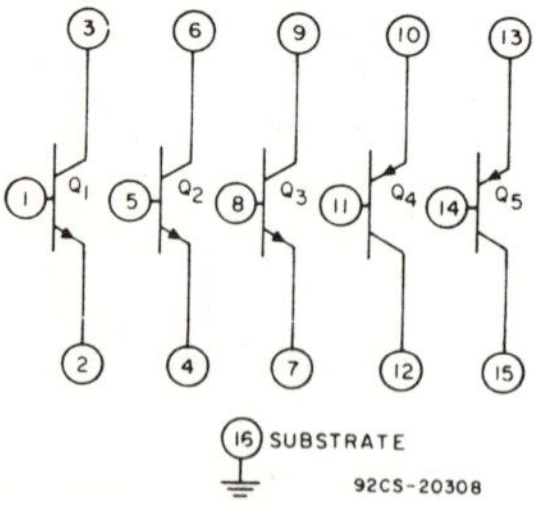

Schematic Diagram

CA3096AE, CA3096E, CA3096CE ESSENTIAL DIFFERENCES

CHARACTERISTIC		CA3096AE	CA3096E	CA3096CE
$V_{(BR)CEO}$ (V) Max.	n-p-n	35	35	24
	p-n-p	–40	–40	–24
$V_{(BR)CBO}$ (V) Max.	n-p-n	45	45	30
	p-n-p	–40	–40	–24
h_{FE} @ 1 mA	n-p-n	150-500	150-500	100-670
	p-n-p	20-150	20-150	15-200
h_{FE} @ 100 μA	p-n-p	40-200	40-200	30-300
I_{CBO} (nA) Max.	n-p-n	40	100	100
	p-n-p	–40	–100	–100
I_{CEO} (nA) Max.	n-p-n	100	1000	1000
	p-n-p	–100	–1000	–1000
$V_{CE}(SAT)$ (V) Max.	p-n-p	0.5	0.7	0.7
$\|V_{IO}\|$ (mW) Max.	n-p-n	5	–	–
	p-n-p	5	–	–
$\|I_{IO}\|$ (μA) Max.	n-p-n	0.6	–	–
	p-n-p	0.25	–	–

Printed in USA/4-77

Supersedes issue dated 10-72

MAXIMUM RATINGS, *Absolute-Maximum Values:*

	EACH N-P-N	EACH P-N-P	
COLLECTOR-TO-EMITTER VOLTAGE, V_{CEO}:			
CA3096AE, CA3096E	35	−40	V
CA3096CE	24	−24	V
COLLECTOR-TO-BASE VOLTAGE, V_{CBO}:			
CA3096AE, CA3096E	45	−40	V
CA3096CE	30	−24	V
COLLECTOR-TO-SUBSTRATE VOLTAGE, V_{CIO}:			
CA3096AE, CA3096E	45	–	V
CA3096CE	30	–	V
EMITTER-TO-SUBSTRATE VOLTAGE, V_{EIO}:			
CA3096AE, CA3096E	–	−40	V
CA3096CE	–	−24	V
EMITTER-TO-BASE VOLTAGE, V_{EBO}:			
CA3096E, CA3096E	6	−40	V
CA3096CE	6	−24	V
COLLECTOR CURRENT, I_C (All Types)	50	−10	mA
POWER DISSIPATION, P_D:			
Up to T_A = 55°C:			
Device (Total)	750		mW
Each Transistor	200		mW
Above T_A = 55°C derate linearly at	6.67		mW/°C
AMBIENT-TEMPERATURE RANGE, T_A:			
Operating		−55 to +125°C	
Storage		−65 to +150°C	
LEAD TEMPERATURE (DURING SOLDERING):			
At distance 1/16 ± 1/32 inch (1.59 ± 0.79 mm) from case for 10 s max.		265°C	

STATIC ELECTRICAL CHARACTERISTICS at T_A = 25°C
For Equipment Design

CHARACTERISTIC	TEST CONDITIONS	LIMITS									UNITS
		CA3096AE			CA3096E			CA3096CE			
		Min.	Typ.	Max.	Min.	Typ.	Max.	Min.	Typ.	Max.	
For Each n-p-n Transistor											
I_{CBO}	V_{CB} = 10 V, I_E = 0	–	0.001	40	–	0.001	100	–	0.001	100	nA
I_{CEO}	V_{CE} = 10 V, I_B = 0	–	0.006	100	–	0.006	1	–	0.006	1	μA
$V_{(BR)CEO}$	I_C = 1 mA, I_B = 0	35	50	–	35	50	–	24	35	–	V
$V_{(BR)CBO}$	I_C = 10 μA, I_E = 0	45	100	–	45	100	–	30	80	–	V
$V_{(BR)CIO}$	I_{CI} = 10 μA, $I_B = I_E$ = 0	45	100	–	45	100	–	30	80	–	V
$V_{(BR)EBO}$	I_E = 10 μA, I_C = 0	6	8	–	6	8	–	6	8	–	V
V_Z	I_Z = 10 μA	6	7.9	9.8	6	7.9	9.8	6	7.9	9.8	V
$V_{CE(SAT)}$	I_C = 10 mA, I_B = 1 mA	–	0.24	0.5	–	0.24	0.7	–	0.24	0.7	V
V_{BE}	I_C = 1 mA, V_{CE} = 5 V	0.6	0.69	0.78	0.6	0.69	0.78	0.6	0.69	0.78	V
h_{FE}		150	390	500	150	390	500	100	390	670	
$\|\Delta V_{BE}/\Delta T\|$	I_C = 1 mA, V_{CE} = 5 V	–	1.9	–	–	1.9	–	–	1.9	–	mV/°C

STATIC ELECTRICAL CHARACTERISTICS at T_A = 25°C (Cont'd)
For Equipment Design

CHARAC-TERISTIC	TEST CONDITIONS	LIMITS									UNITS
		CA3096AE			CA3096E			CA3096CE			
		Min.	Typ.	Max.	Min.	Typ.	Max.	Min.	Typ.	Max.	
For Each p-n-p Transistor											
I_{CBO}	$V_{CB} = -10$ V, $I_E = 0$	–	−0.006	– 40	–	−0.06	−100	–	−0.06	-100	nA
I_{CEO}	$V_{CE} = -10$V, $I_B = 0$	–	−0.12	– 100	–	−0.12	– 1	–	−0.12	– 1	μA
$V_{(BR)CEO}$	$I_C = -100\,\mu$A, $I_B = 0$	−40	−75	–	−40	−75	–	−24	−30	–	V
$V_{(BR)CBO}$	$I_C = -10\,\mu$A, $I_E = 0$	−40	−80	–	−40	−80	–	−24	−60	–	V
$V_{(BR)EBO}$	$I_E = -10\,\mu$A, $I_C = 0$	−40	−100	–	−40	−100	–	−24	−80	–	V
$V_{(BR)EIO}$	$I_{EI} = 10\,\mu$A, $I_B = I_C = 0$	−40	−100	–	−40	−100	–	−24	−80	–	V
$V_{CE(SAT)}$	$I_C = -1$ mA, $I_B = -100\,\mu$A	–	−0.16	−0.4	–	−0.16	−0.4	–	−0.16	−0.4	V
V_{BE}	$I_C = -100\,\mu$A, $V_{CE} = -5$ V	−0.5	−0.6	−0.7	−0.5	−0.6	−0.7	−0.5	−0.6	−0.7	V
h_{FE}	$I_C = -100\,\mu$A, $V_{CE} = -5$ V	40	85	200	40	85	200	30	85	300	
	$I_C = -1$ mA, $V_{CE} = -5$ V	20	47	150	20	47	150	15	47	200	
$\|\Delta V_{BE}/\Delta T\|$	$I_C = -100\,\mu$A, $V_{CE} = -5$ V	–	2.2	–	–	2.2	–	–	2.2	–	mV/°C

I_{CBO} Collector-Cutoff Current

I_{CEO} Collector-Cutoff Current

$V_{(BR)CEO}$ Collector-to-Emitter Breakdown Voltage

$V_{(BR)CBO}$ Collector-to-Base Breakdown Voltage

$V_{(BR)CIO}$ Collector-to-Substrate Breakdown Voltage

$V_{(BR)EBO}$ Emitter-to-Base Breakdown Voltage

V_Z Emitter-to-Base Zener Voltage

$V_{CE(SAT)}$ Collector-to-Emitter Saturation Voltage

V_{BE} Base-to-Emitter Voltage

h_{FE} DC Forward-Current Transfer Ratio

$|\Delta V_{BE}/\Delta T|$ Magnitude of Temperature Coefficient: (for each transistor)

STATIC ELECTRICAL CHARACTERISTICS at T_A = 25°C (CA3096AE Only)
For Equipment Design

CHARACTERISTIC	TEST CONDITIONS	LIMITS CA3096AE Min.	Typ.	Max.	UNITS
For Transistors Q1 and Q2 (as a Differential Amplifier)					
Absolute Input Offset Voltage, $\lvert V_{IO} \rvert$	V_{CE} = 5 V, I_C = 1 mA	–	0.3	5	mV
Absolute Input Offset Current, $\lvert I_{IO} \rvert$		–	0.07	0.6	μA
Absolute Input Offset Voltage Temperature Coefficient, $\frac{\lvert \Delta V_{IO} \rvert}{\Delta T}$		–	1.1	–	μV/°C
For Transistors Q4 and Q5 (As a Differential Amplifier)					
Absolute Input Offset Voltage, $\lvert V_{IO} \rvert$	V_{CE} = –5 V, I_C = –100 μA R_S = 0	–	0.15	5	mV
Absolute Input Offset Current, $\lvert I_{IO} \rvert$		–	2	250	nA
Absolute Input Offset Voltage Temperature Coefficient, $\frac{\lvert \Delta V_{IO} \rvert}{\Delta T}$		–	0.54	–	μV/°C

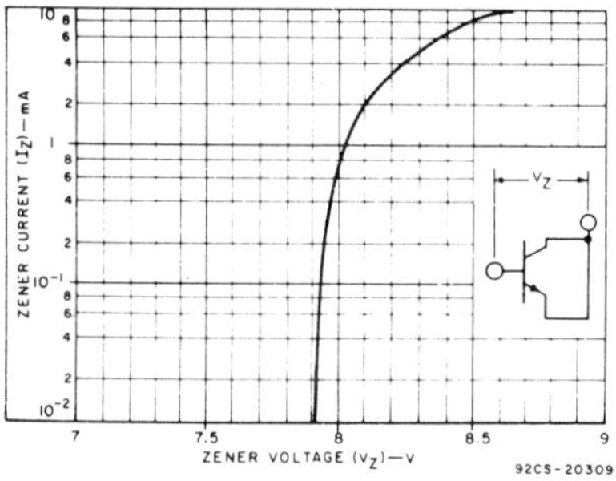

Fig. 1 – Base-to-emitter zener characteristic (n-p-n).

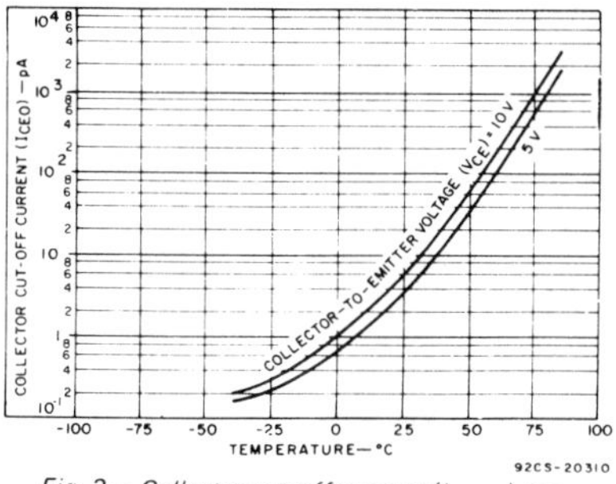

Fig. 2 – Collector cut-off current (I_{CEO}) as a function of temperature (n-p-n).

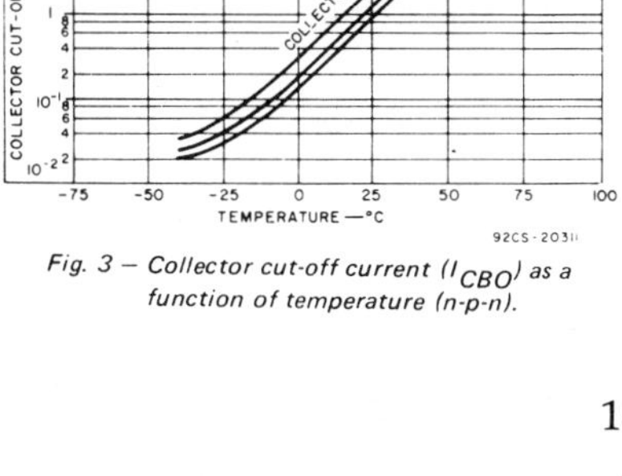

Fig. 3 – Collector cut-off current (I_{CBO}) as a function of temperature (n-p-n).

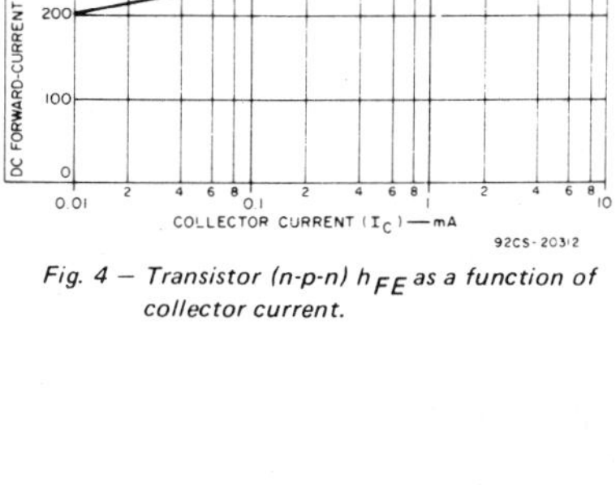

Fig. 4 – Transistor (n-p-n) h_{FE} as a function of collector current.

DYNAMIC

ELECTRICAL CHARACTERISTICS at $T_A = 25°C$

Typical Values Intended Only for Design Guidance

CHARACTERISTICS	TEST CONDITIONS	TYPICAL VALUES	UNITS
For Each n-p-n Transistor			
Noise Figure (low frequency), NF	$f = 1\,kHz$, $V_{CE} = 5\,V$, $I_C = 1\,mA$, $R_S = 1\,k\Omega$	2.2	dB
Low-Frequency, Input Resistance, R_i	$f = 1.0\,kHz$, $V_{CE} = 5\,V$, $I_C = 1\,mA$	10	kΩ
Low-Frequency Output Resistance, R_o		80	kΩ
Admittance Characteristics:	$f = 1\,MHz$, $V_{CE} = 5\,V$, $I_C = 1\,mA$		
Forward Transfer Admittance, y_{fe} — g_{fe}		7.5	mmho
Forward Transfer Admittance, y_{fe} — b_{fe}		−j13	
Input Admittance, y_{ie} — g_{ie}		2.2	mmho
Input Admittance, y_{ie} — b_{ie}		j3.1	
Output Admittance, y_{oe} — g_{oe}		0.76	mmho
Output Admittance, y_{oe} — b_{oe}		j2.4	
Gain-Bandwidth Product, f_T	$V_{CE} = 5\,V$, $I_C = 1.0\,mA$	280	MHz
	$V_{CE} = 5\,V$, $I_C = 5\,mA$	335	
Emitter-to-Base Capacitance, C_{EB}	$V_{EB} = 3\,V$	0.75	pF
Collector-to-Base Capacitance, C_{CB}	$V_{CB} = 3\,V$	0.46	pF
Collector-to-Substrate Capacitance, C_{CI}	$V_{CI} = 3\,V$	3.2	pF
For Each p-n-p Transistor			
Noise Figure (low frequency), NF	$f = 1\,kHz$, $I_C = 100\,\mu A$, $R_S = 1\,k\Omega$	3	dB
Low-Frequency Input Resistance, R_i	$f = 1\,kHz$, $V_{CE} = 5\,V$, $I_C = 100\,\mu A$	27	kΩ
Low-Frequency Output Resistance, R_o		680	kΩ
Gain-Bandwidth Product, f_T	$V_{CE} = 5\,V$, $I_C = 100\,\mu A$	6.8	MHz
Emitter-to-Base Capacitance, C_{EB}	$V_{EB} = -3\,V$	0.85	pF
Collector-to-Base Capacitance, C_{CB}	$V_{CB} = -3\,V$	2.25	pF
Base-to-Substrate Capacitance, C_{BI}	$V_{BI} = 3\,V$	3.05	pF

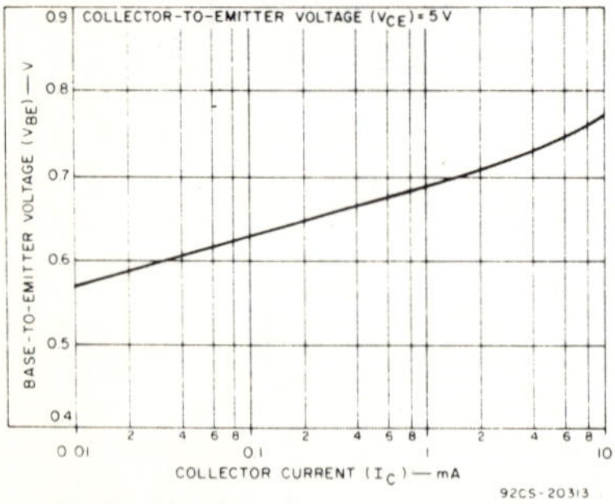

Fig. 5 – V_{BE} (n-p-n) as a function of collector current.

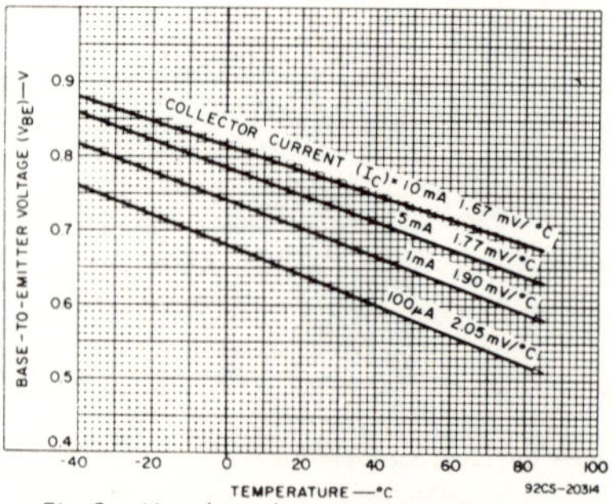

Fig. 6 – V_{BE} (n-p-n) as a function of temperature.

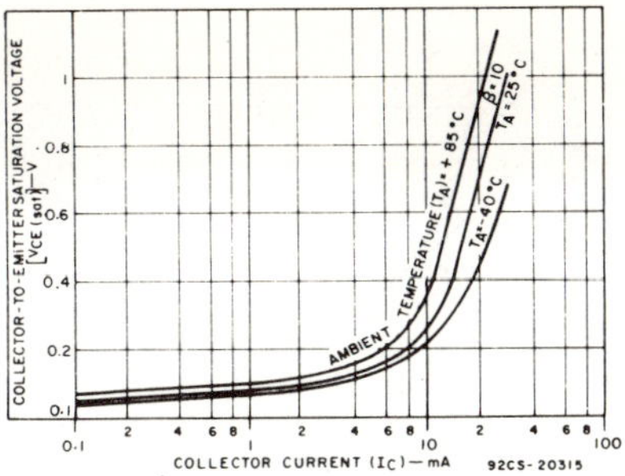

Fig. 7 – V_{CE}(SAT) (n-p-n) as a function of collector current.

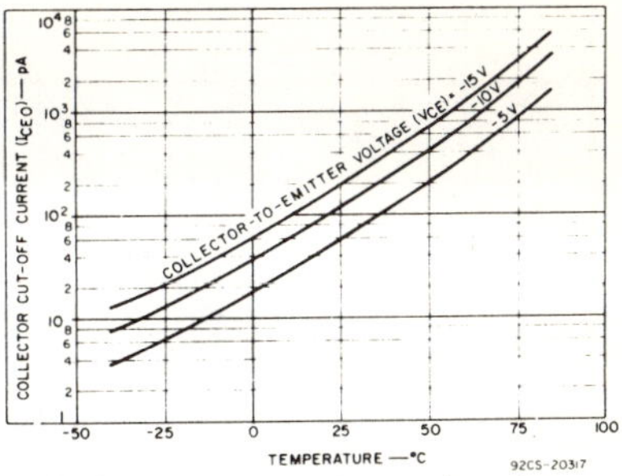

Fig. 8 – Collector cut-off current (I_{CEO}) as a function of temperature (p-n-p).

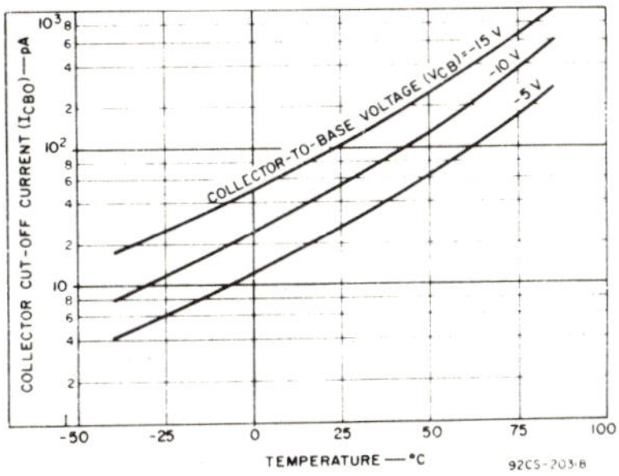

Fig. 9 – Collector cut-off current (I_{CBO}) as a function of temperature (p-n-p).

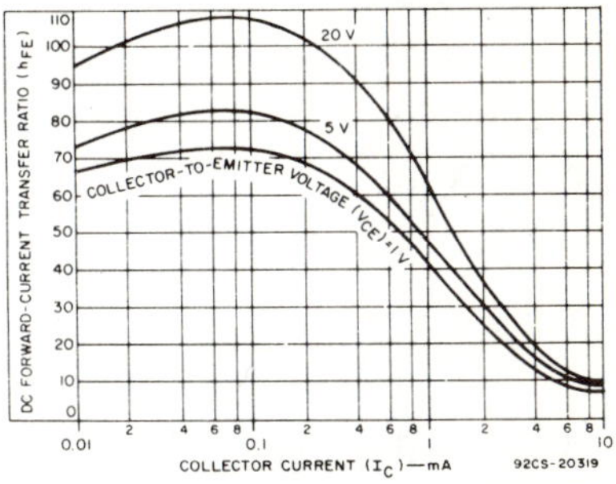

Fig. 10 – Transistor (p-n-p) h_{FE} as a function of collector current.

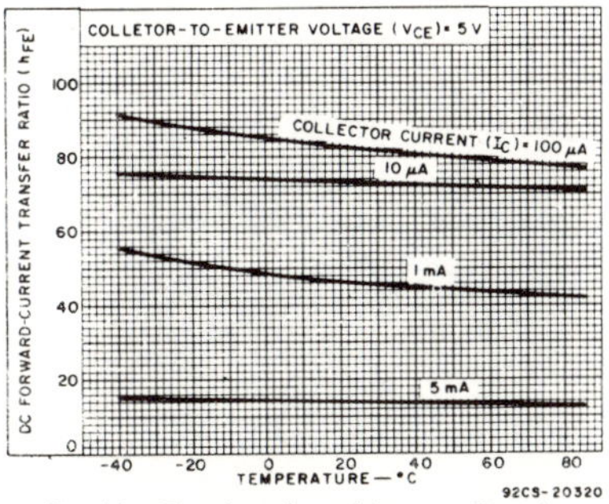

Fig. 11 – Transistor (p-n-p) h_{FE} as a function of temperature.

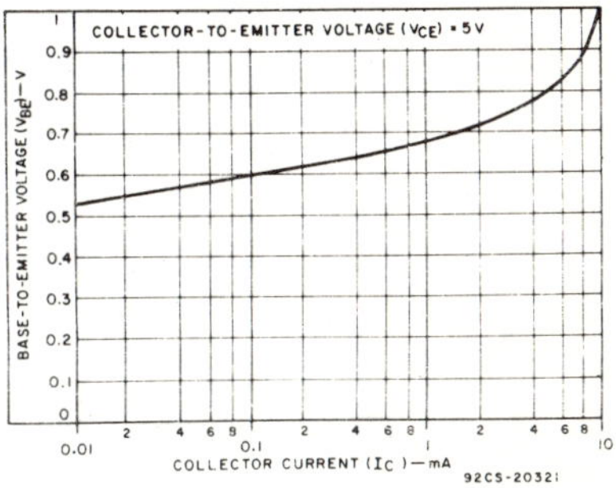

Fig. 12 – V_{BE} (p-n-p) as a function of collector current.

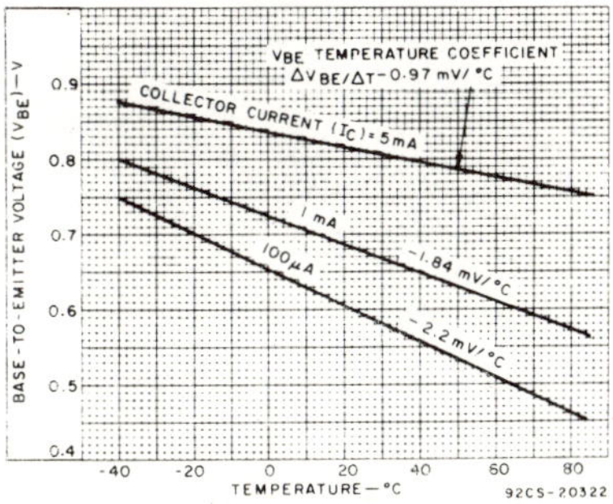

Fig. 13 – V_{BE} (p-n-p) as a function of temperature.

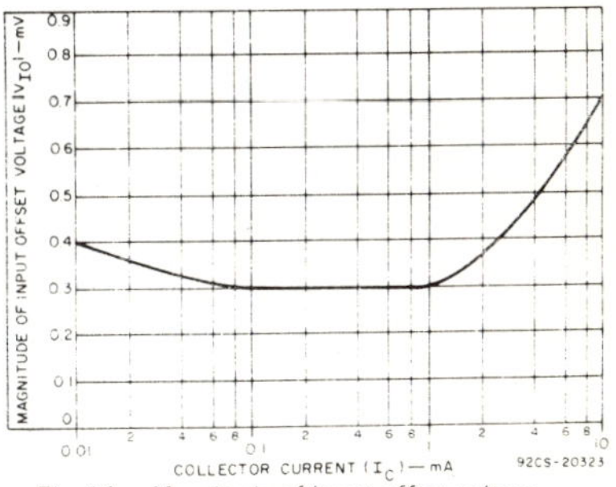

Fig. 14 – Magnitude of input offset voltage $|V_{IO}|$ as a function of collector current for n-p-n transistor Q_1–Q_2.

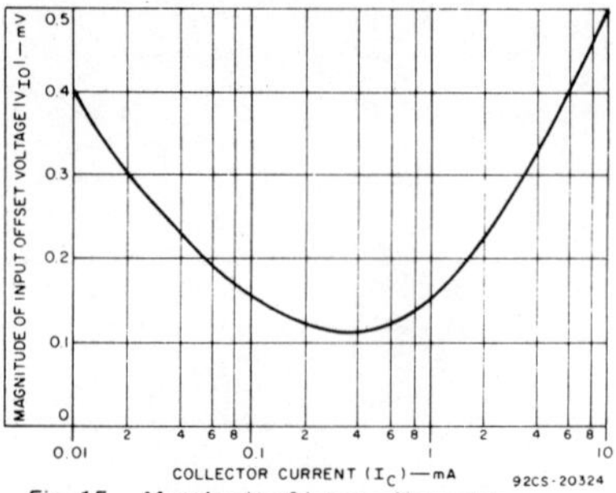

Fig. 15 – Magnitude of input offset voltage $|V_{IO}|$ as a function of collector current for p-n-p transistor Q_4–Q_5.

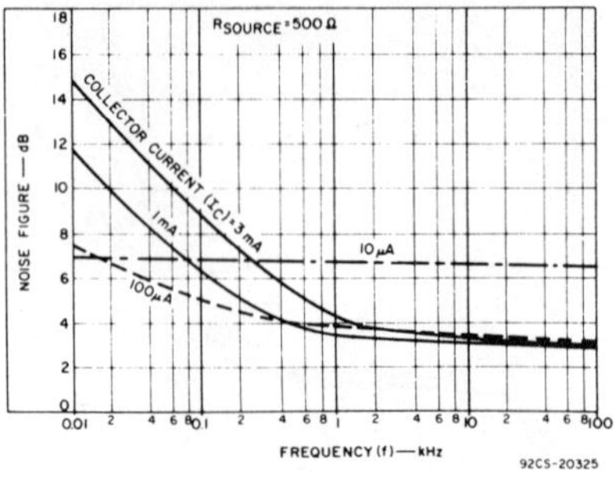

Fig. 16 – Noise figure as a function of frequency for n-p-n transistors.

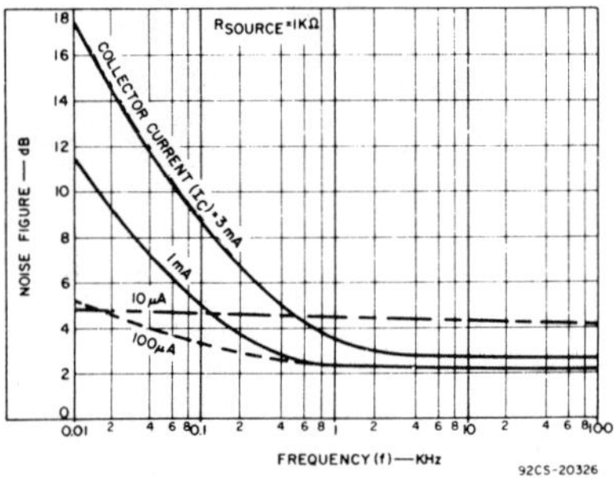

Fig. 17 – Noise figure as a function of frequency for n-p-n transistors.

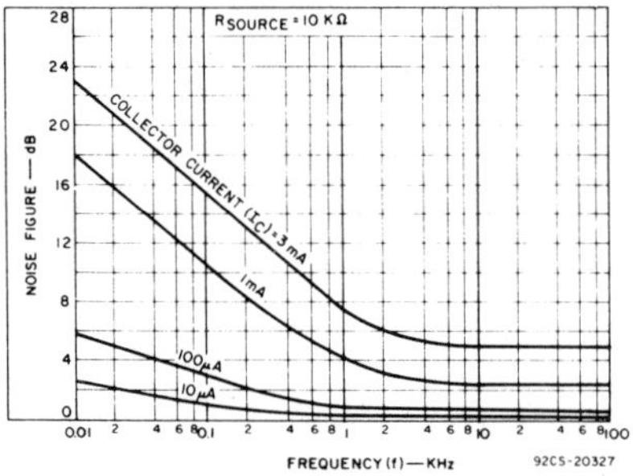

Fig. 18 – Noise as a function of frequency for n-p-n transistors.

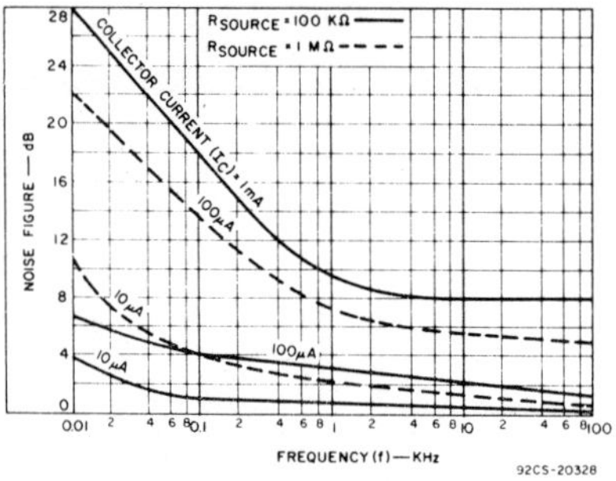

Fig. 19 – Noise figure as a function of frequency for n-p-n transistors.

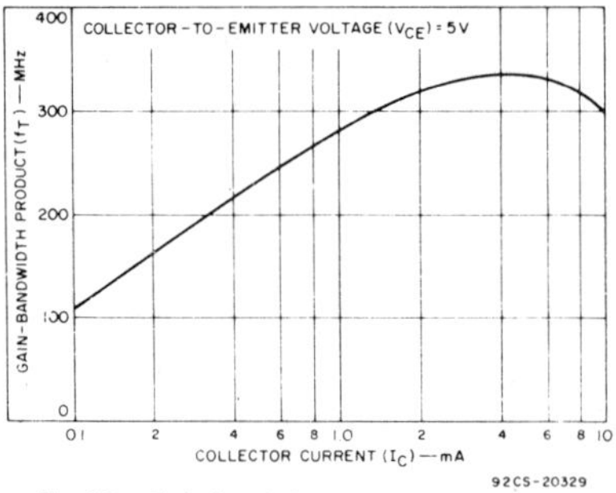

Fig. 20 – Gain-bandwidth product as a function of collector current (n-p-n).

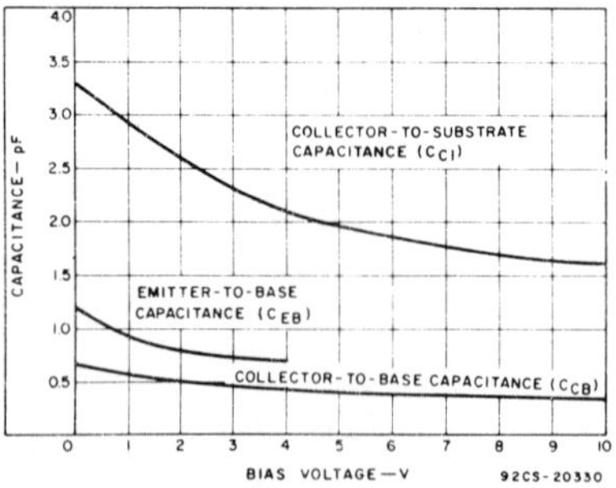

Fig. 21 – Capacitance as a function of bias voltage (n-p-n).

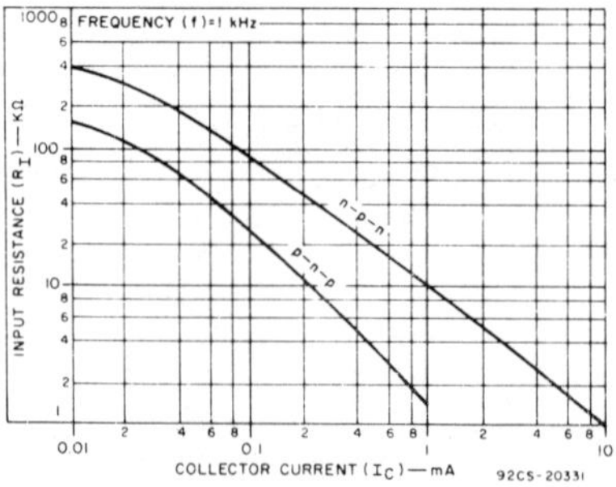

Fig. 22 – Input resistance as a function of collector current.

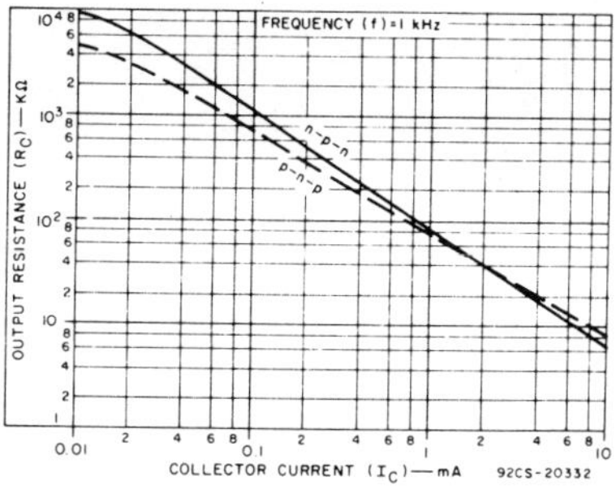

Fig. 23 – Output resistance as a function of collector current.

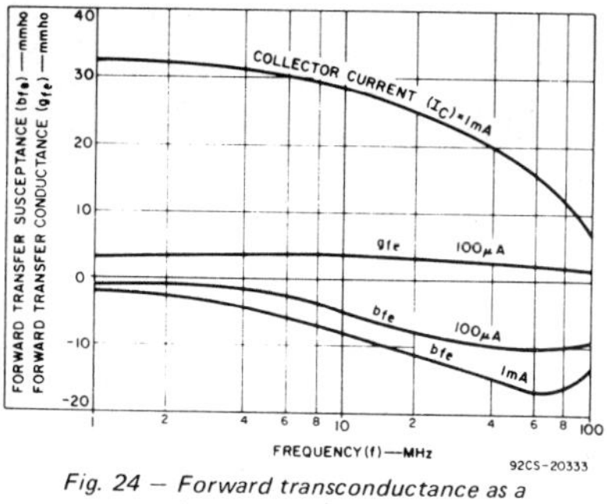

Fig. 24 – Forward transconductance as a function of frequency.

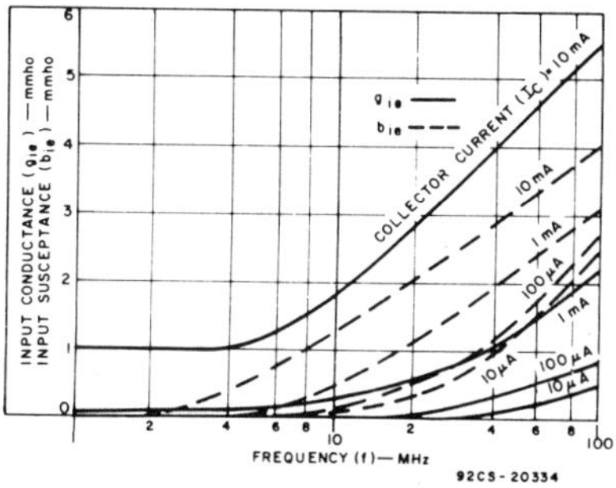

Fig. 25 – Input admittance as a function of frequency.

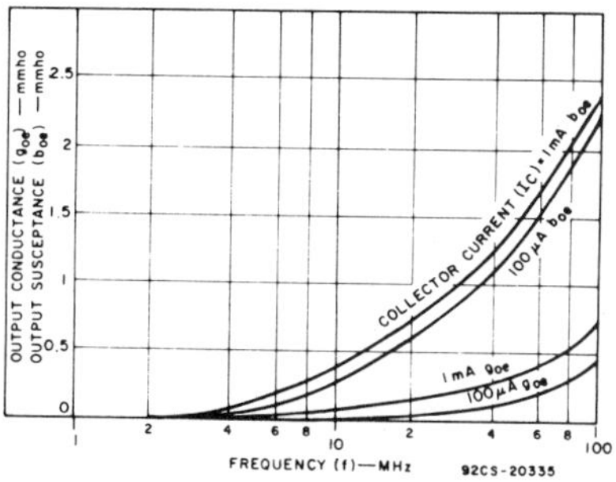

Fig. 26 – Output admittance as a function of frequency.

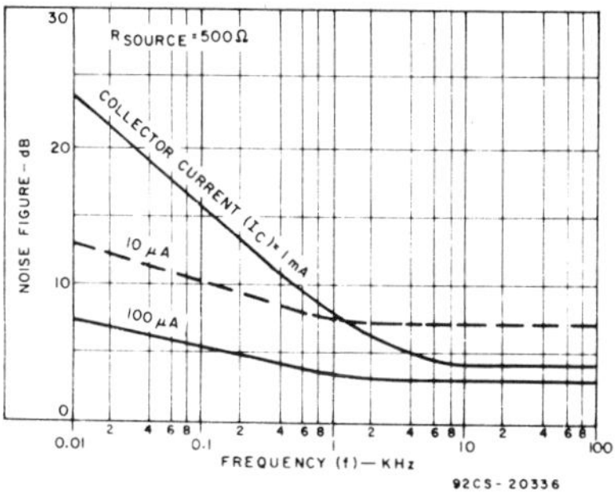

Fig. 27 – Noise figure as a function of frequency (p-n-p).

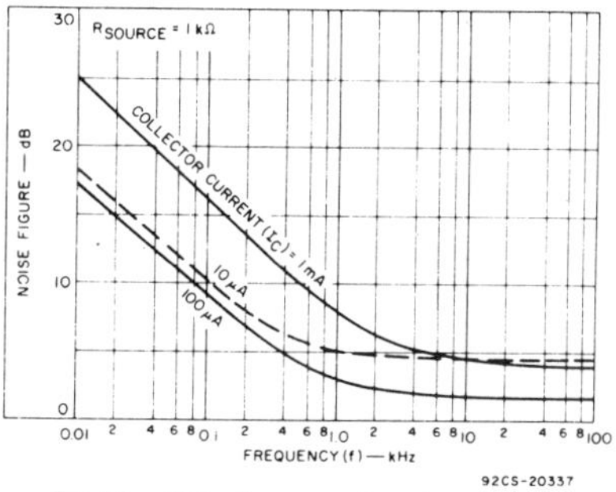

Fig. 28 – Noise figure as a function of frequency (p-n-p).

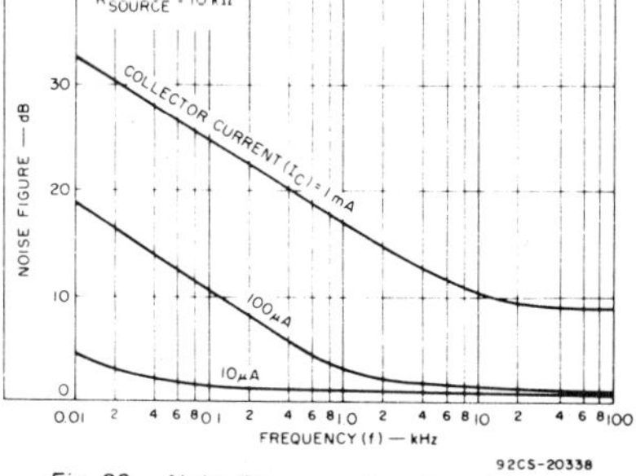

Fig. 29 – Noise figure as a function of frequency (p-n-p).

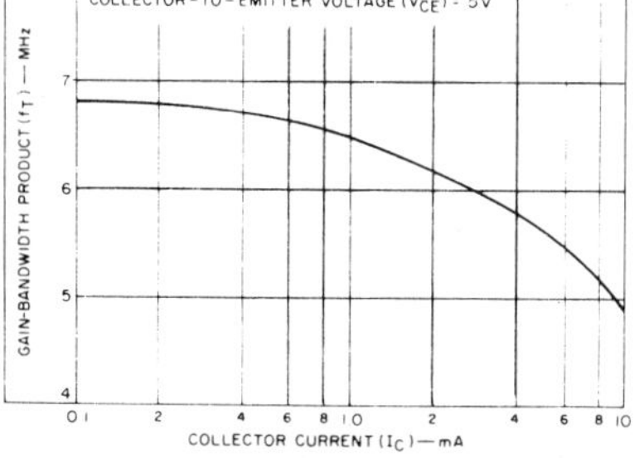

Fig. 30 – Gain-bandwidth product as a function of collector current (p-n-p).

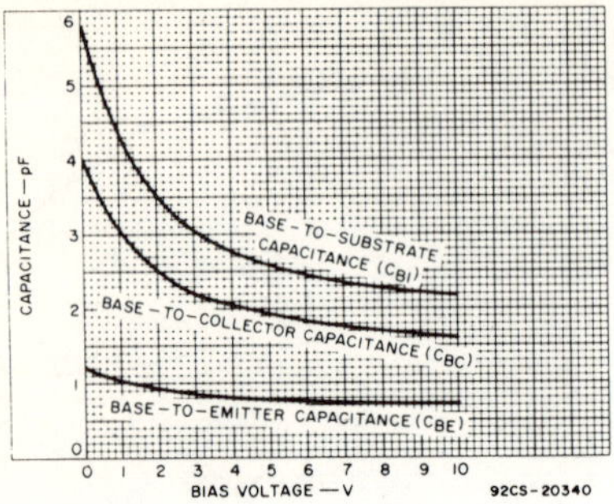

Fig. 31 – Capacitance as a function of bias voltage (p-n-p).

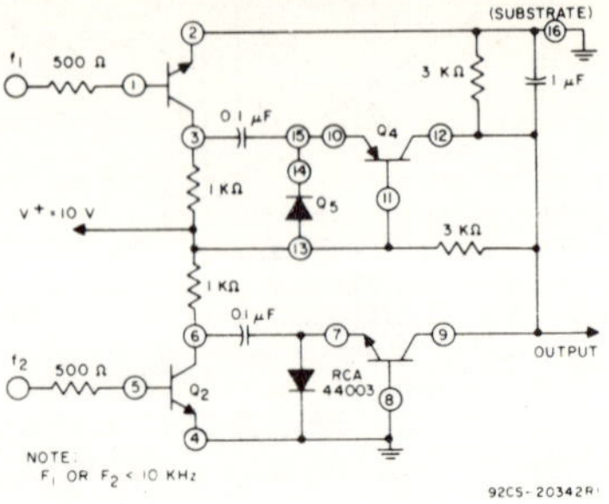

Fig. 32 – Frequency comparator using CA3096E.

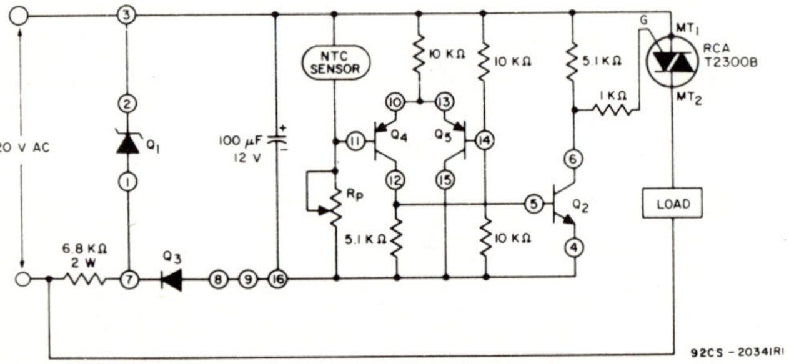

Fig. 33 – Line-operated level switch using CA3096AE or CA3096E.

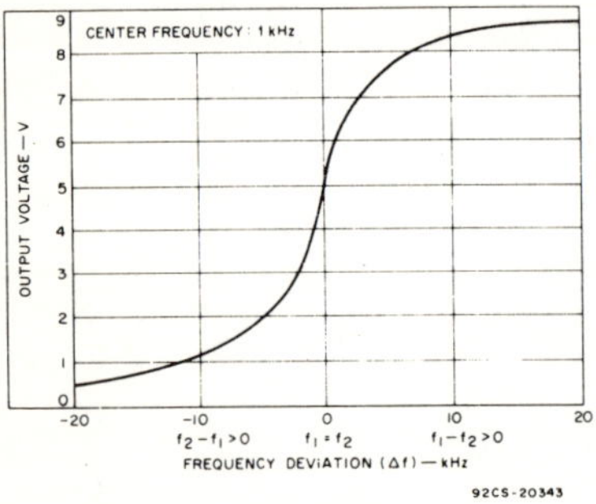

Fig. 34 – Frequency comparator characteristics.

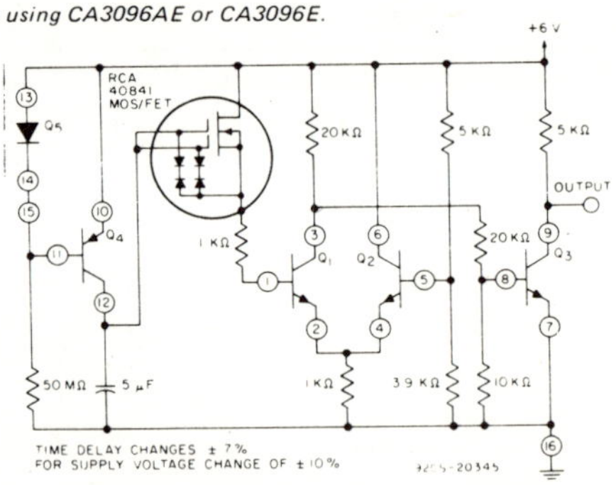

Fig. 35 – One-minute timer using CA3096AE and a MOS/FET.

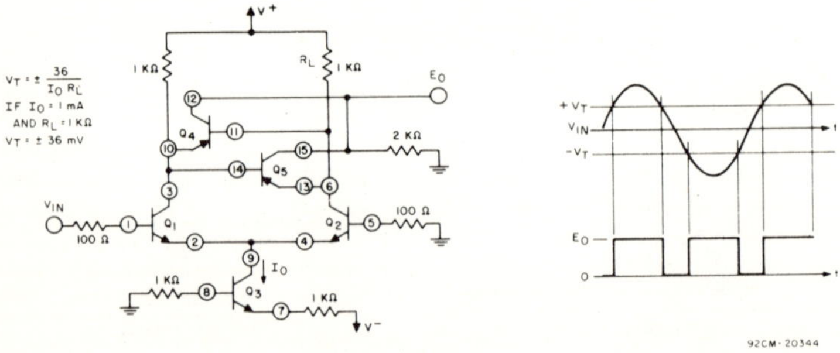

Fig. 36 – CA3096AE small-signal zero-voltage detector having noise immunity.

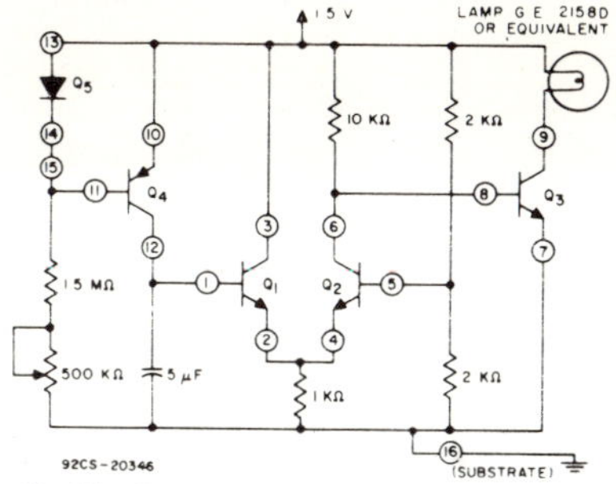

Fig. 37 – Ten-second timer operated form 1.5-volt supply using CA3096E.

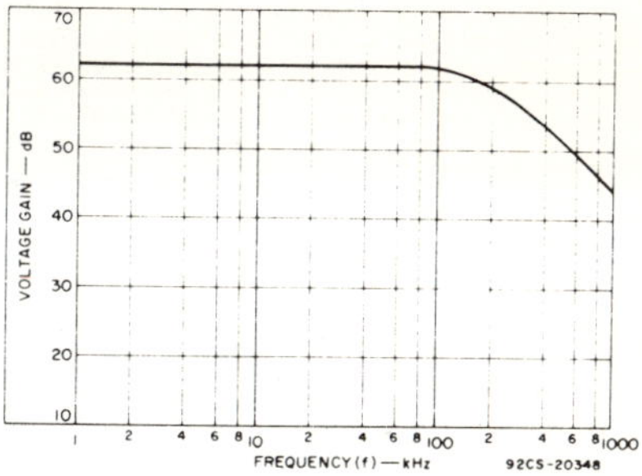

Fig. 38 – Gain-frequency characteristics.

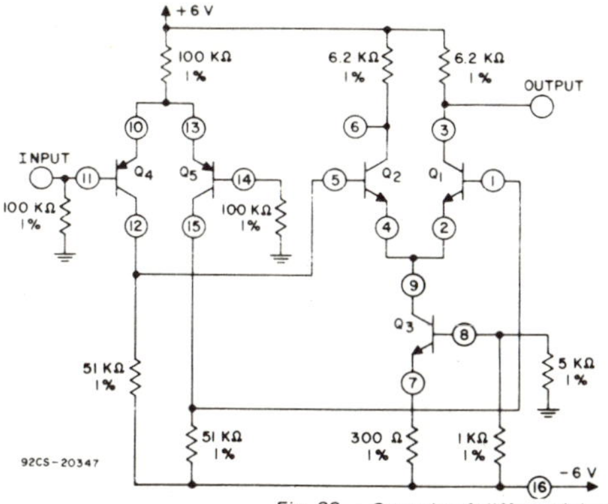

Fig. 39 – Cascade of differential amplifiers using CA3096AE.

Features:

1. Can be operated with either dual supply or single supply.
2. Wide-input common-mode range +5 V to −5 V.
3. Low bias current: <1 μA.

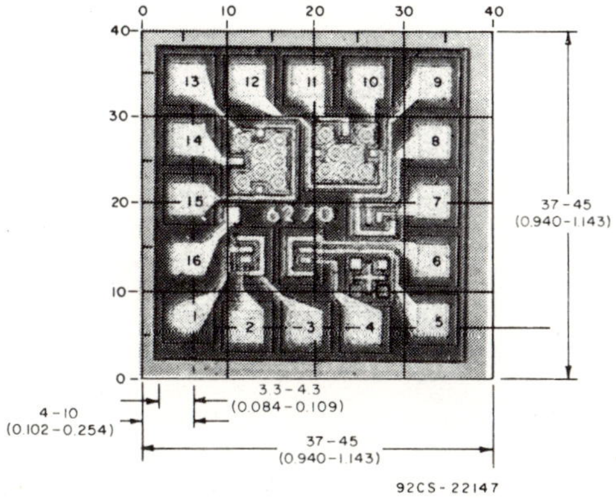

CA3096CH

Dimensions in parentheses are in millimeters and are derived from the basic inch dimensions as indicated. Grid graduations are in mils (10^{-3} inch).

The photographs and dimensions represent a chip when it is part of the wafer. When the wafer is cut into chips, the cleavage angles are 57° instead of 90° with respect to the face of the chip. Therefore, the isolated chip is actually 7 mils (0.17 mm) larger in both dimensions.

DIMENSIONAL OUTLINE
16-LEAD DUAL-IN-PLASTIC PACKAGE
JEDEC MO-001-AC

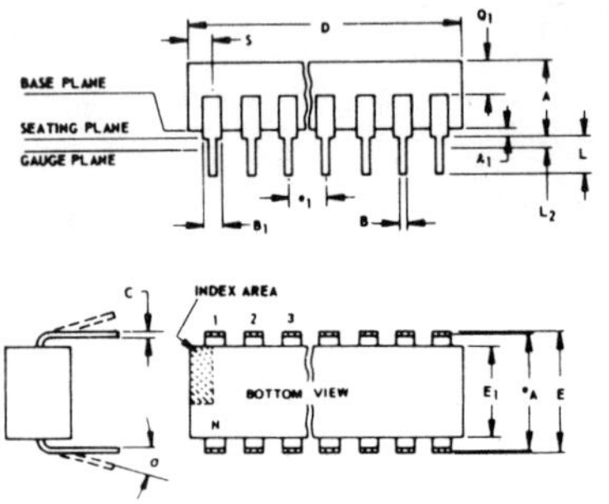

SYMBOL	INCHES		NOTE	MILLIMETERS	
	MIN.	MAX.		MIN.	MAX.
A	0.155	0.200		3.94	5.08
A_1	0.020	0.050		0.51	1.27
B	0.014	0.020		0.356	0.508
B_1	0.035	0.065		0.89	1.65
C	0.008	0.012	1	0.204	0.304
D	0.745	0.785		18.93	19.93
E	0.300	0.325		7.62	8.25
E_1	0.240	0.260		6.10	6.60
e_1	0.100 TP		2	2.54 TP	
e_A	0.300 TP		2, 3	7.62 TP	
L	0.125	0.150		3.18	3.81
L_2	0.000	0.030		0.000	0.76
α	0°	15°	4	0°	15°
N	16		5	16	
N_1	0		6	0	
Q_1	0.040	0.075		1.02	1.90
S	0.015	0.060		0.39	1.52

92CM-15967R4

NOTES:

Refer to Rules for Dimensioning (JEDEC Publication No. 95) for Axial Lead Product Outlines.

1. When this device is supplied solder-dipped, the maximum lead thickness (narrow portion) will not exceed 0.013" (0.33 mm).
2. Leads within 0.005" (0.12 mm) radius of True Position (TP) at gauge plane with maximum material condition and unit installed.
3. e_A applies in zone L_2 when unit installed.
4. α applies to spread leads prior to installation.
5. N is the maximum quantity of lead positions.
6. N_1 is the quantity of allowable missing leads.

When incorporating RCA Solid State Devices in equipment, it is recommended that the designer refer to "Operating Considerations for RCA Solid State Devices", Form No. 1CE-402, available on request from RCA Solid State Division, Box 3200, Somerville, N. J. 08876.

RCA Solid State Division

Linear Integrated Circuits

Monolithic Silicon

CA3084

File Number 482

14-Lead Dual-In-Line Plastic Package

General-Purpose P-N-P Transistor Array

FEATURES

- Matched transistor pair (Q1 and Q2)
  - V_{IO} (V_{BE} matched): ± 6mV max.
  - I_{IO} (at 100 μA): ± 0.6 μA
- Wide operating current range
- Low noise figure - - 3.2 dB typ. at 1 kHz

APPLICATIONS

- General use in signal processing systems having low-power and low-frequency requirements
- Differential amplifiers
- Temperature compensated amplifiers
- Active loads for differential amplifiers using n-p-n transistors
- Complementary uses with RCA n-p-n transistor arrays

RCA-CA3084* is a general-purpose silicon p-n-p transistor array incorporating two independent transistors, a Darlington circuit, and a current-mirror pair with a shared diode.

The two independent transistors in the array may be used in a variety of circuit applications. The Darlington pair may be employed as the equivalent of a single high-beta transistor. The current-mirror pair is well suited for constant-current applications and can also be used as the active loads in a differential amplifier which uses n-p-n transistors.

The total array is especially useful for a wide range of applications in systems having low-power and low-frequency requirements. Although the transistors may be used as discrete units in conventional circuits, they offer the advantages inherent in integrated-circuit construction, that is, to provide close electrical and thermal matching.

The CA3084 utilizes the 14-lead dual-in-line plastic package.

*Formerly developmental type TA5799A.

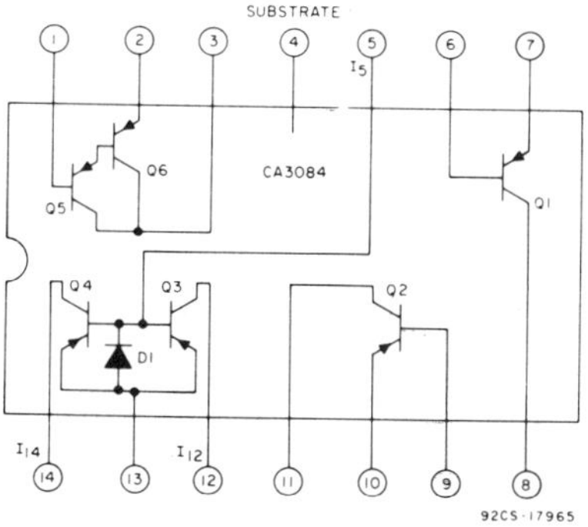

Fig. 1 – Functional diagram of the CA3084.

CA3084 General-Purpose P-N-P Transistor Array

Printed in USA/3-71

File No.482 CA3084

ELECTRICAL CHARACTERISTICS at T_A = 25°C
For Equipment Design

CHARACTERISTICS	SYMBOL	TEST CONDITIONS	Typ. Characteristics Curve Fig. No.	LIMITS Min.	Typ.	Max.	UNITS
For Each Transistor:							
Collector-Cutoff Current	I_{CBO}	V_{CB} = −10V, I_E = 0	2	–	−0.055	−100	nA
Collector-Cutoff Current	I_{CEO}	V_{CE} = −10V, I_B = 0	3	–	−0.12	−100	nA
Collector-to-Emitter Breakdown Voltage	$V_{(BR)CEO}$	I_{CE} = −100μA, I_B = 0	–	−40	−70	–	V
Collector-to-Base Breakdown Voltage	$V_{(BR)CBO}$	I_{CB} = −100μA, I_E = 0	–	−40	−80	–	V
Emitter-to-Base Breakdown Voltage	$V_{(BR)EBO}$	I_{EB} = −100μA, I_C = 0	–	−40	−100	–	V
Emitter-to-Substrate Breakdown Voltage	$V_{(BR)EIO}$	I_{EI} = 100μA	–	−40	−100	–	V
Collector-to-Emitter Saturation Voltage	V_{CEsat}	I_E = 1mA, I_B = 100μA	4	–	−0.125	−0.25	V
Base-to-Emitter Voltage	V_{BE}	I_E = 100μA, V_{CE} = −10V	5	−0.50	−0.59	−0.68	V
DC Forward-Current Transfer Ratio	h_{FE}		7	15	40	–	
For Transistors Q1 and Q2 (As a Differential Amplifier):							
Magnitude of Input Offset Voltage	$\lvert V_{IO} \rvert$	I_E = 100μA, V_{CE} = −10V	8	–	0.422	6	mV
Input Offset Current	I_{IO}		–	−0.6	0	0.6	μA
For Transistors Q3 and Q4 (Current-Mirror Configuration):							
Collector Current	I_C	V_{CE} = −10V, V_{CIO} = −10V,	10	0.85	1.00	1.15	μA
Magnitude of Collector Current Ratio	$\lvert I_C(Q3)/I_C(Q4) \rvert$	Term. 13 = Gnd. I_5 = −100μA,	11	0.90	1.00	1.10	
For Transistors Q5 and Q6 (Darlington Configuration):							
Collector-Cutoff Current	I_{CEO}	V_{CE} = −10V, I_B = 0	–	–	–	−1.0	μA
Base-to-Emitter Voltage	V_{BE}	I_E = 100μA, V_{CE} = −10V	13	0.92	1.07	1.20	V
DC Forward-Current Transfer Ratio	h_{FE}		15	100	1230	–	

ELECTRICAL CHARACTERISTICS at T_A = 25°C
Typical Values Intended Only For Design Guidance

Magnitude of Temperature Coefficient:							
V_{BE} (for each transistor)	$\lvert \Delta V_{BE}/\Delta T \rvert$	I_E = 100μA, V_{CE} = −10V	6		−1.78		mV/°C
V_{IO} (as a differential amplifier)	$\lvert \Delta V_{IO}/\Delta T \rvert$		9		0.54		μV/°C
V_{BE} (Darlington configuration)	$\lvert \Delta V_{BE}/\Delta T \rvert$		14		−3.7		mV/°C
For Each Transistor:							
Input Resistan e	R_I	f = 1kHz, V_{CE} = −10V, I_C = −100μA	19		9		kΩ
Output Resistance	R_O		20	–	600	–	kΩ
Forward Transconductance	g_m		22	–	3	–	mmho
Collector-to-Base Capacitance	C_{CBO}	I_{CB} = 0	23	–	3.3	–	pF
Collector-to-Emitter Capacitance	C_{CEO}	I_{CE} = 0	23	–	2.5	–	pF
Base-to-Substrate Capacitance	C_{BIO}	I_{CIO} = 0	23	–	4.5	–	pF

MAXIMUM RATINGS, Absolute-Maximum Values at T_A = 25°C

Dissipation:		
Any one transistor	200	mW
Total package	750	mW
Above T_A = 55°C	derate linearly6.67	mW/°C
Ambient Temperature Range:		
Operating	−40 to +85	°C
Storage	−55 to +150	°C
The following ratings apply for each transistor in the device:		
Collector-to-Emitter Voltage (V_{CEO})	−40	V
Collector-to-Base Voltage (V_{CBO})	−40	V
Base-to-Substrate Voltage (V_{BIO})*	−40	V
Emitter-to-Base Voltage (V_{EBO})	−40	V
Collector Current (I_C)	−10	mA

*The base of each transistor of the CA3084 is isolated from the substrate by an integral diode. The substrate must be connected to a voltage which is more negative than any base voltage in order to maintain isolation between transistors and provide normal transistor action. To avoid undesired coupling between transistors, the substrate terminal (4) should be maintained at either DC or signal (AC) ground. A suitable bypass capacitor can be used to establish a signal ground.

STATIC CHARACTERISTICS FOR EACH TRANSISTOR

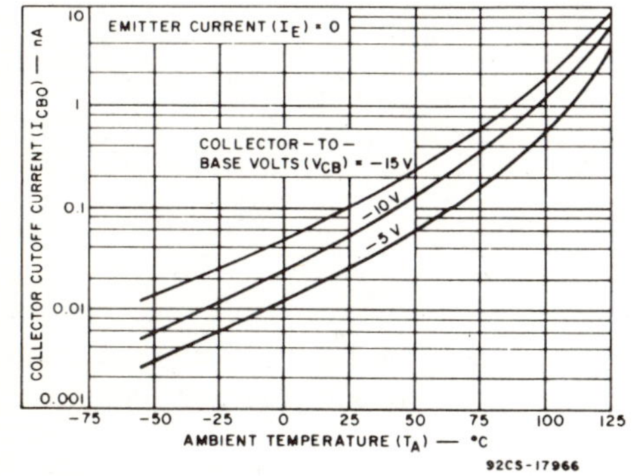

Fig.2 – I_{CBO} vs T_A.

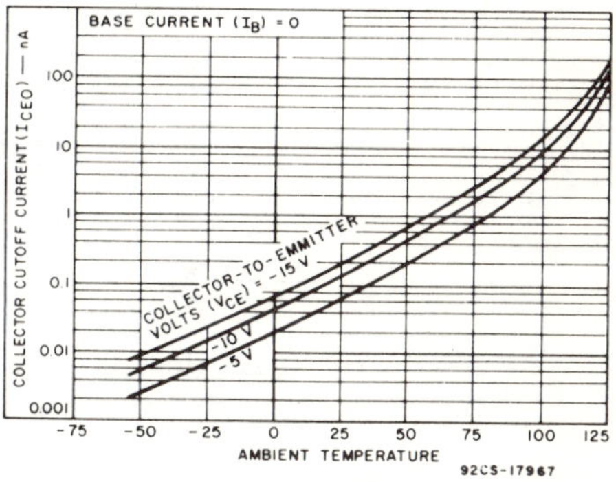

Fig.3 – I_{CEO} vs T_A.

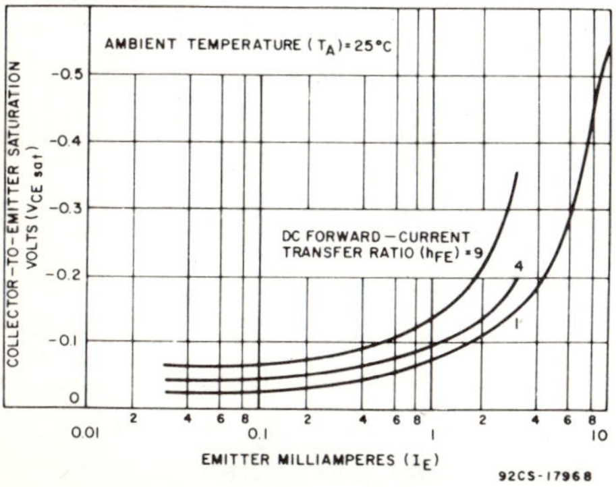

Fig.4 – V_{CEsat} vs I_E.

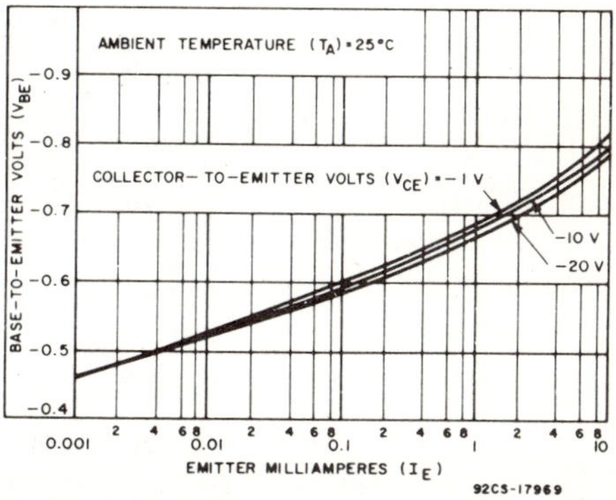

Fig.5 – V_{BE} vs I_E.

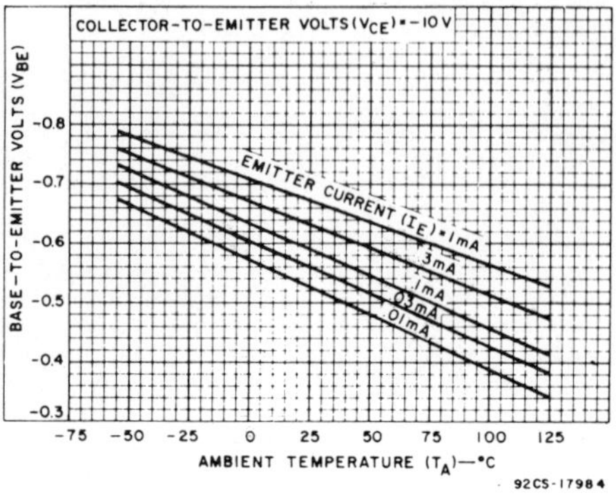

Fig.6— V_{BE} vs T_A •

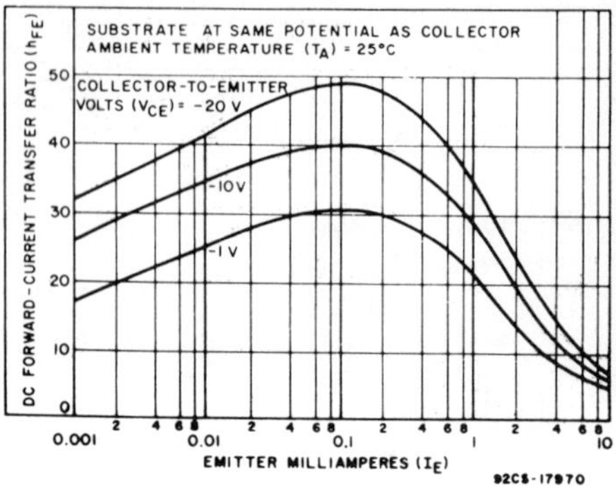

Fig.7— h_{FE} vs I_E •

STATIC CHARACTERISTICS FOR DIFFERENTIAL AMPLIFIER

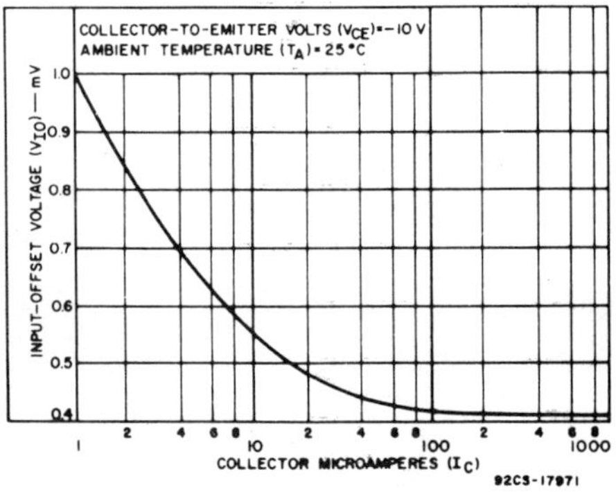

Fig.8—V_{IO} vs I_C, (transistors Q1 and Q2 as a differential amplifier).

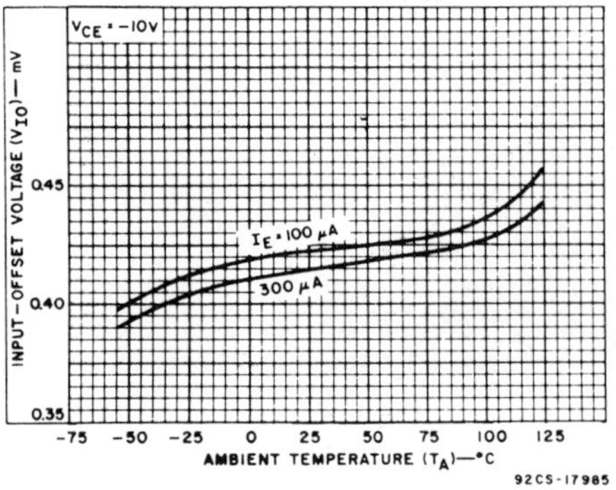

Fig.9—V_{IO} vs T_A (transistors Q1 and Q2 as a differential amplifier).

STATIC CHARACTERISTICS FOR CURRENT-MIRROR CONFIGURATION

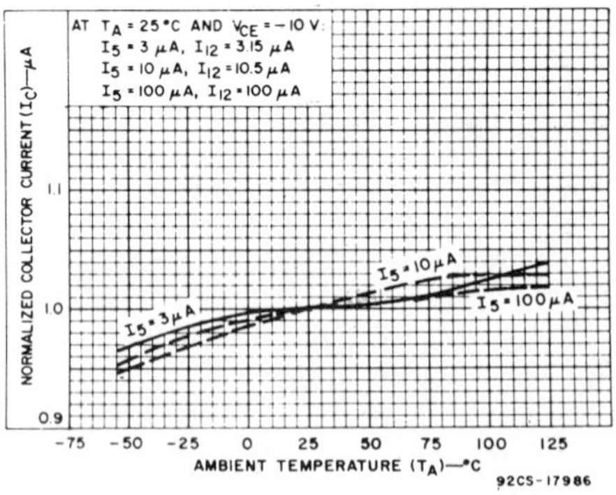

Fig.10—Normalized I_C vs T_A (transistors Q3 and Q4 in a current-mirror configuration.

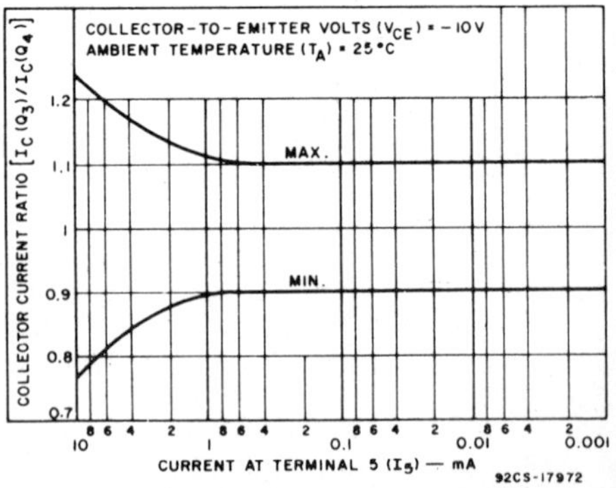

Fig.11—I_C ratio vs I_5 (transistors Q3 and Q4 in a current-mirror configuration.

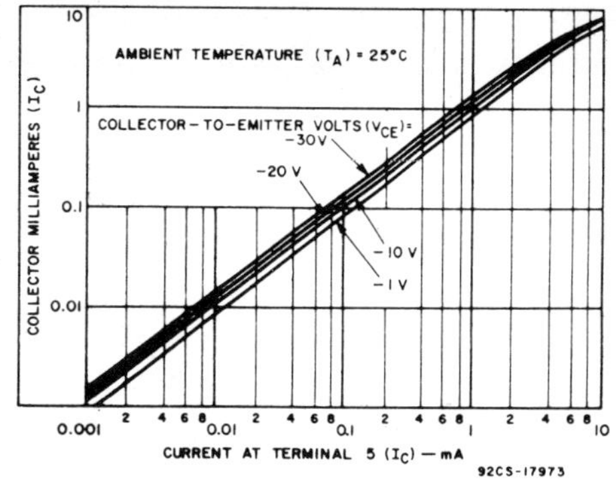

Fig.12—I_C vs I_5 (transistors Q3 and Q4 in a current-mirror configuration).

STATIC CHARACTERISTICS FOR DARLINGTON CONFIGURATION

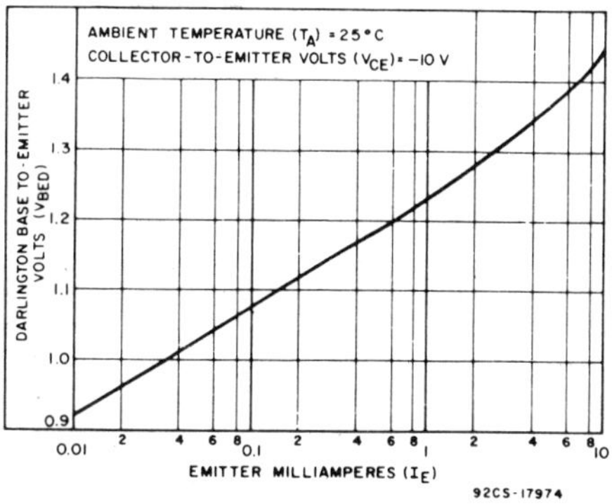

Fig.13—V_{BE} vs I_E (transistors Q5 and Q6 in a darlington configuration).

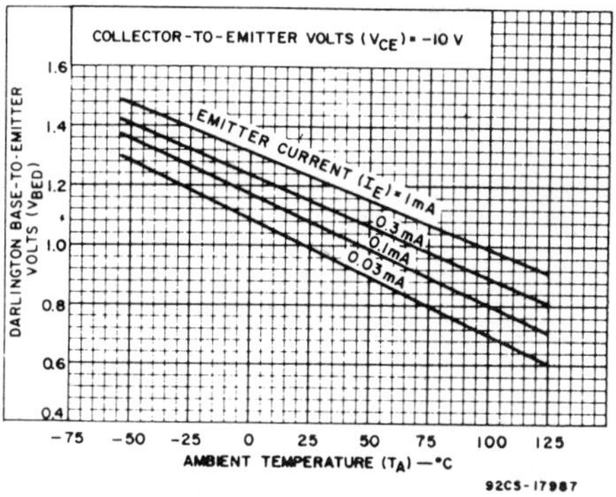

Fig.14—V_{BE} vs T_A (transistors Q5 and Q6 in a darlington configuration).

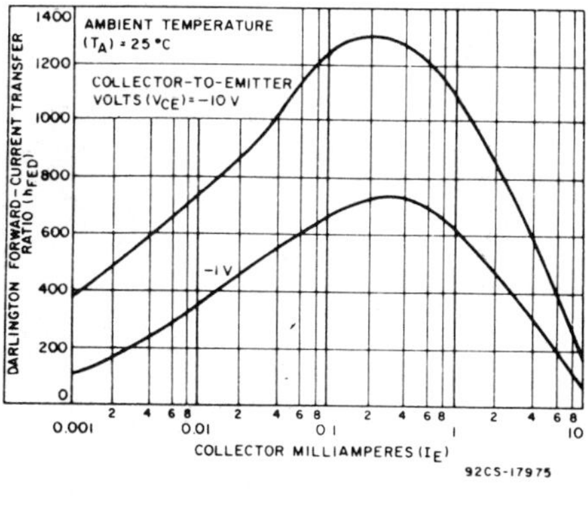

Fig.15—h_{FE} vs I_E (transistors Q5 and Q6 in a darlington configuration).

DYNAMIC CHARACTERISTICS FOR EACH TRANSISTOR

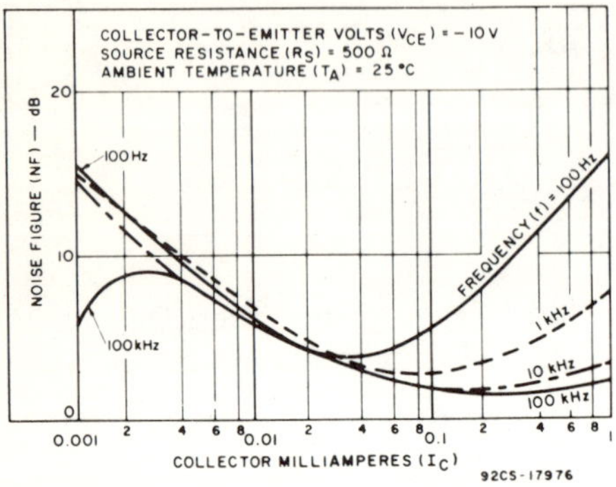

Fig.16 – NF vs I_C at R_S = 500Ω

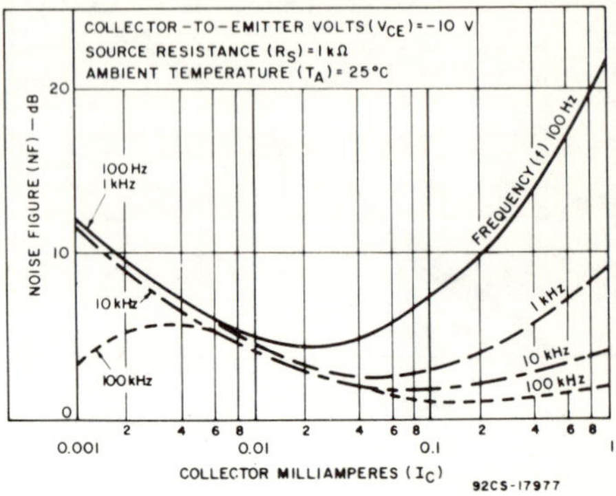

Fig.17 – NF vs I_C at R_S = 1kΩ

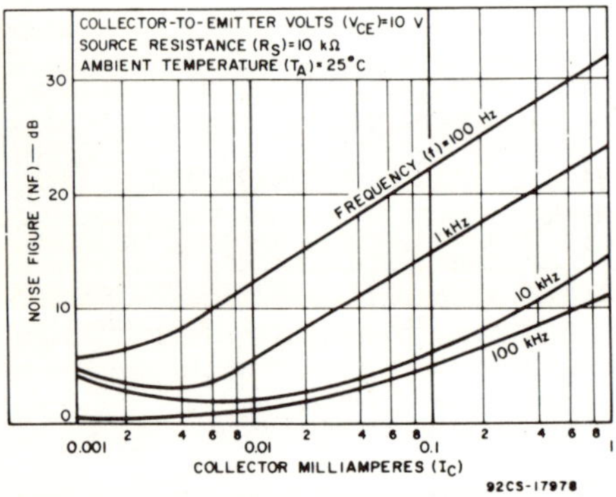

Fig.18 – NF vs I_C at R_S = 10kΩ

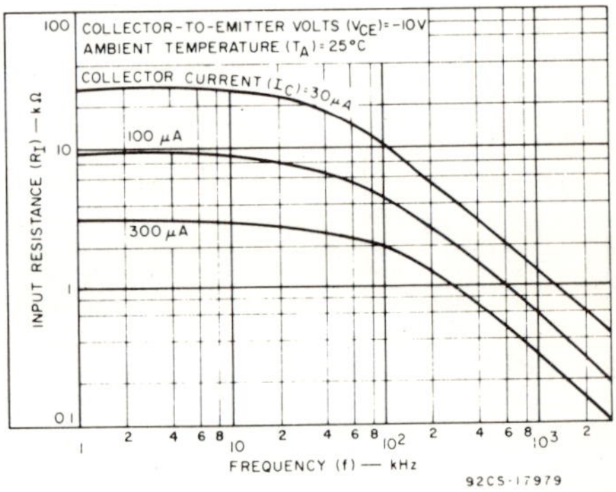

Fig.19 – R_I vs f

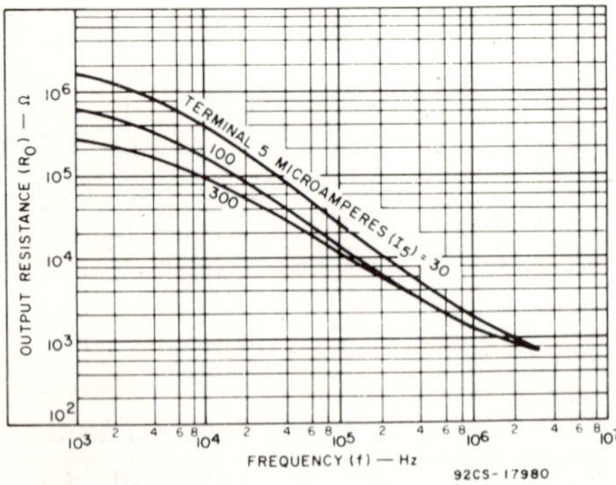

Fig.20 – R_O vs f

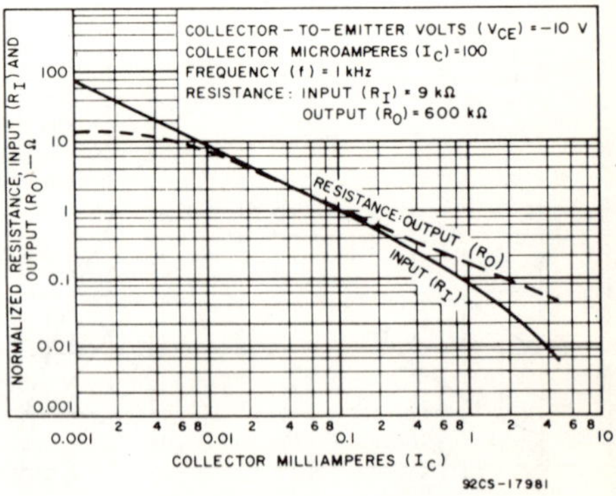

Fig.21 – Normalized R_I and R_O vs I_C

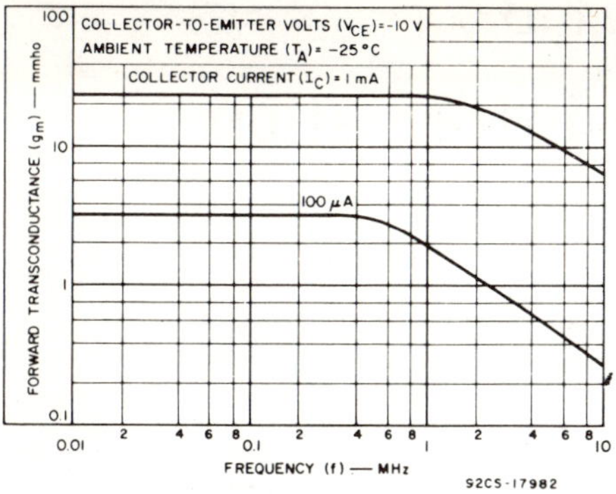

Fig.22 – g_m vs f

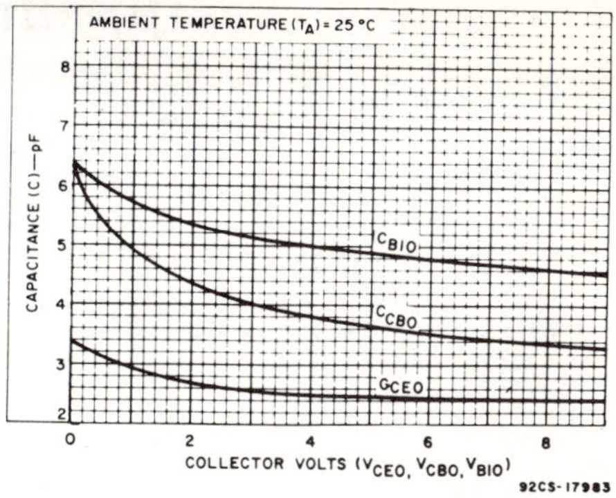

Fig.23 – Transistor capacitances vs collector voltages (V_{CEO}, V_{CBO}, V_{CIO})

DIMENSIONAL OUTLINE

14-LEAD DUAL-IN-LINE PLASTIC PACKAGE–JEDEC MO-001-AB

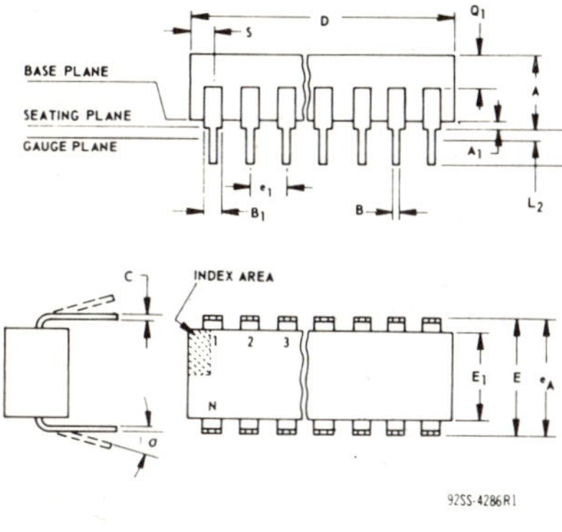

SYMBOL	INCHES MIN.	INCHES MAX.	NOTE	MILLIMETERS MIN.	MILLIMETERS MAX.
A	0.155	0.200		3.94	5.08
A_1	0.020	0.050		0.51	1.27
B	0.014	0.020		0.356	0.508
B_1	0.050	0.065		1.27	1.65
C	0.008	0.012		0.204	0.304
D	0.745	0.770		18.93	19.55
E	0.300	0.325		7.62	8.25
E_1	0.240	0.260		6.10	6.60
e_1	0.100 TP		2	2.54 TP	
e_A	0.300 TP		2, 3	7.62 TP	
L	0.125	0.150		3.18	3.81
L_2	0.000	0.030		0.000	0.76
α	0°	15°	4	0°	15°
N	14		5	14	
N_1	0		6	0	
Q_1	0.040	0.075		1.02	1.90
S	0.065	0.090		1.66	2.28

NOTES:

1. Refer to Rules for Dimensioning Axial Lead Product Outlines.
2. Leads within 0.005" (0.12 mm) radius of True Position (TP) at guage plane with maximum material condition and unit installed.
3. e_A applies in zone L_2 when unit installed.
4. α applies to spread leads prior to installation.
5. N is the maximum quantity of lead positions.
6. N_1 is the quantity of allowable missing leads.

When incorporating RCA Solid State Devices in equipment, it is recommended that the designer refer to "Operating Considerations for RCA Solid State Devices", (Form No. 1CE-402), available on request from RCA Commercial Engineering, Harrison, N. J. 07029.

Appendix B

IC Array Source Information

Source	*Device*
Exar Integrated Systems 750 Palomar Ave. Sunnyvale, Calif. 94086	Semi-custom IC's (inquire direct)
Fairchild 464 Ellis St. Mountain View, Calif. 94042	μA3018/μA3018A, μA3019 μA3026/μA3054, μA3036, μA3039, μA3045/μA3046/ μA3086
Interdesign 1255 Reamwood Ave. Sunnyvale, Calif. 94086	Semi-custom IC's (inquire direct)
Lithic Systems 15800 Sanborn Rd. P.O. Box 478 Saratoga, Calif. 95070	LS3018, LS3046/LS3086
Motorola Semiconductor Products 5005 East McDowell Rd. Phoenix, Ariz., 85008	MC3346/(CA3046), MC3386/ (CA3086), CA3054
National Semiconductor Corp. 2900 Semiconductor Drive Santa Clara, Calif. 95051	LM3018/LM3018A, LM3019 LM3026, LM3039, LM3045/LM3046/ LM3086, LM3118/LM3118A, LM3146/LM3146A
Plessey Semiconductors 1674 McGaw Ave. Santa Ana, Calif. 92705	SL3018/SL3018A, SL3045/ SL3046/SL3086, SL3145/SL3146 SL3118/SL3118A, SL3083/ SL3183, SL3093
RCA Solid State Division Route 202 Somerville, N.J. 08876	CA3018/CA3018A, CA3019 CA3026/CA3054, CA3039, CA3045/ CA3046/CA3086, CA3049/CA3102

CA3060, CA3080, CA3083, CA3093, CA3095, CA3097, CA3127, CA3118/CA3118A, CA3146/CA3146A, CA3183/ CA3183A, CA3138/CA3724/ CA3725, CA3141

Silicon General Inc. 7382 Bolsa Ave. Westminster, Calif. 92683	SG3018/SG3018A, SG3146 SG3186
Sprague Electric Co. Semiconductor Division 115 N.E. Cutoff Worcester, Mass. 01606	ULS2045A/ULN-2046A/ ULN2086A, ULN2081A/ULN2082A

Appendix C

Mail Order Distributor Listing

Virtually all semiconductor manufacturers market ICs through a nationwide market of distributors, most of which cater primarily to the OEM buyer. As a rule, these distributors do not encourage relatively small volume mail orders from individuals, as is typical of the home experimenter and hobbyist.

The following list of distributors are specifically organized to handle such orders and can accommodate the mail order buyer of ICs and related components and hardware.

Active Electronic Sales
Po. Box 1035
Framingham, Mass. 01701

Circuit Specialists
P.O. Box 3047
Scottsdale, Ariz. 85257

International Electronics Unlimited
P.O. Box 449
Carmel Valley, Calif. 93924

Quest
P.O. Box 4430S
Santa Clara, Calif. 95054

Digi-Key
P.O. Box 677
Thief River Falls, Minn. 56701

Index